T0362185

A HISTORY OF EARLY TELEVISION

ROUTLEDGE LIBRARY OF MEDIA AND CULTURAL STUDIES

A HISTORY OF EARLY TELEVISION

Volume III

Selected and with a new introduction by
Stephen Herbert

Routledge
Taylor & Francis Group

LONDON AND NEW YORK

First published 2004
by Routledge

2 Park Square, Milton Park, Abingdon, Oxfordshire OX14 4RN
52 Vanderbilt Avenue, New York, NY 10017

Routledge is an imprint of the Taylor & Francis Group, an informa business

First issued in hardback 2019

Editorial matter and selection © 2004 Stephen Herbert;
individual owners retain copyright in their own material

Typeset in Times by Keystroke, Jacaranda Lodge, Wolverhampton

All rights reserved. No part of this book may be reprinted or reproduced or utilised
in any form or by any electronic, mechanical, or other means, now known or
hereafter invented, including photocopying and recording, or in any information
storage or retrieval system, without permission in writing from the publishers.

Notice:
Product or corporate names may be trademarks or registered trademarks,
and are used only for identification and explanation without intent to
infringe.

British Library Cataloguing in Publication Data
A catalogue record for this book is available from the British Library

Library of Congress Cataloging in Publication Data
A catalog record for this book has been requested

ISBN 13: 978-0-415-32665-0 (Set)
ISBN 13: 978-0-415-32668-1 (Volume III) (hbk)

ISBN 13: 978-1-032-66039-4 (Volume III pbk)

Publisher's note
The publisher has gone to great lengths to ensure the quality of this reprint but
points out that some imperfections in the original book may be apparent.

CONTENTS

ACKNOWLEDGEMENTS

The publishers would like to thank the following for permission to reprint their material:

The New York Times for permission to reprint:
[The Tide of Events] 'Telefilmed Faces', Orrin E. Dunlap Jr, *New York Times*, 20 September 1936.
[The Tide of Events] 'Television called answer to song how to keep 'em down on the farm', *New York Times*, 20 September 1936.
'Television in store carries hat styles', *New York Times*, 27 April 1939.
'Telecast of President at the World's Fair To Start Wheels of New Industry', Orin E. Dunlap Jr., *New York Times*, 30 April 1939.
'Television Now Drops Mantle of Mystery And Public Becomes Its Judge', *New York Times*, 30 April 1939.

Popular Mechanics for permission to reprint 'Where is Television Now?', *Popular Mechanics*, August 1938.

David Sarnoff Library, Princeton, New Jersey for permission to reprint:
David Sarnoff, 'RCA's Development of Television', typescript in RCA Annual Meeting Minutes and Proceedings, 7 April 1936.
David Sarnoff, 'Statement on Television', RCA press release, 7 May 1935.
David Sarnoff, 'Television in Advertising', excerpted from a stencil copy of 'The Message of Radio', addressed to Advertising Federation of America, 29 June 1936.

HarperCollins for permission to reprint *Television – A Struggle for Power*, Frank C. Waldrop and Joseph Borkin (New York: William Morrow and Company, 1938), pp. 3–21, 70–80, 121–130, 216–240, 254–269, 271–273.

David Sarnoff Library, Princeton, New Jersey, and the Journal of Applied Physics for permission to reprint 'Probable Influences of Television on Society', David Sarnoff, *Journal of Applied Physics*, July 1939.

Norton and Company for permission to reprint *We Present Television*, John Porterfield and Kay Reynolds, eds (New York: W.W. Norton & Company, 1940).

Disclaimer

The publishers have made every effort to contact authors/copyright holders of works reprinted in *A History of Early Television*. This has not been possible in every case, however, and we would welcome correspondence from those individuals/companies whom we have been unable to trace.

INTRODUCTION

Part 5: A New Era
High-definition and regular broadcasting in the United States of America

Through the mid-1930s in the USA, transmissions from experimental stations continued. American newspapers and magazines commented on various aspects of the new medium during those years. [The Tide of Events] 'Telefilmed Faces', *New York Times* (20 September 1936) reports that President Roosevelt had been broadcast on TV in New York (apparently from news film shots) as part of a daily but irregular schedule. In the same issue, 'Television called answer to song how to keep 'em down on the farm' muses on whether seeing the world on television will keep the rural population entertained at home, or will they want to travel to the places they have seen on the television screen. But ten years after Baird's experimental trans-Atlantic transmission, there was still no major public service, and the American periodical *Popular Mechanics* noted: 'People still are asking, "When will we have television?"'. 'Where is Television Now?' (August 1938) explains to the magazine's general readership the problems of high-definition broadcasting – the necessary short waves used have very limited range – the difficulties of image dissection and reconstitution, and the lack of standardisation.

RCA's television guru David Sarnoff is represented in *Television. Collected addresses and papers on the future of the new art and its recent technical developments* (1936–46), a collection of papers published by the RCA Institute's Technical Press. Some are reproduced here: 'RCA's Development of Television'; 'Television'; and 'Television in Advertising' (all by David Sarnoff, reproduced here in their original typescript form); 'Commercial Television – and its needs', by Alfred N. Goldsmith; 'Television and the Electron', by Vladimir Kosma Zworykin; and 'Television Studio Technic', by Albert W. Protzman.

Our next section comprises extracts from *Television – A Struggle for Power*, by Frank C. Waldrop and Joseph Borkin, published in 1938: a book with a strident tone, mostly concerned with the wider problems of broadcasting in general, and from which I have selected the chapters most concerned with television. Political and economic factors are the main concerns.

The 1939 New York World's Fair was the venue and occasion chosen for the major public launch of television in America, and was well covered by the *New York Times*. And it wasn't just broadcasting. Department store Bloomingdale's installed viewing 'kinets' (monitors) throughout the store for showing live demonstrations of hat fashions, advertised as 'History by Television', and reported as 'Television in store carries hat styles' (27 April 1939). An advertisement by Davega City Radio stores the next day, 'Hear, See, Television', promoted RCA Victor and Du Mont receivers – and the RCA Victor radio adapted for use with

1

television attachment – tempting potential viewers with the information that 'This Sunday ... President Roosevelt will officially open the N. Y. World's Fair and, at the same time, a new industry TELEVISION will be launched'. Du Mont's own advertisement for TV receivers in the same day's paper took a whole page: 'Television Gives its "Coming-out Party" Sunday'.

'Radio Straps on its Camera and goes to the Fair' was the banner headline in the Radio section of the *New York Times* (30 April 1939), with pieces covering 'Telecast of President at the World's Fair To Start Wheels of New Industry': 'There may be 200 tele-radios in operation within a fifty-mile radius of Manhattan's spires, but there is no gauge or turnstile in the air, as at the Fair, to clock the number of sightseers. Possibly 500 will look in on the opening when the curtain rolls up ... ' and 'Television Now Drops Mantle of Mystery And Public Becomes Its Judge' indicates that there had been 'no rush on the part of New Yorkers ... to have video [television] sets installed ... '. The public was waiting to see whether TV caught on – and of course most of the country was still out of range. In 'Probable Influences of Television on Society', *Journal of Applied Physics* (July 1939), David Sarnoff anticipates the future of the new medium.

Our final facsimile reproduction is another complete book: *We Present Television,* edited by John Porterfield and Kay Reynolds (New York: W.W. Norton & Co., 1940); a comprehensive coverage of the subject to date, from an American perspective, with reflections on potential future developments. Important chapters include those concerning programming, the special techniques required of director and actor, and the problems of finance. The first chapter notes:

> The National Broadcasting Company inaugurated its first high-definition service, under technical standards laid down by the radio industry, with the transmission of the opening day ceremonies at the New York World's Fair of 1939. Since that time it has maintained a regular program schedule for pioneer viewers within the area, extending roughly sixty miles in all directions from the Empire State Building, covered by NBC's New York City transmitter. The Don Lee Broadcasting System has also begun a service in Los Angeles under the same technical standards, and the General Electric Company has completed its transmitter near Schenectady for serving the Albany–Troy–Schenectady area. The GE station plans regular relays of New York NBC programs over its service area, thus marking the beginning of the first television network.

The final chapter, 'The Challenge of Television', concludes: 'Television is only waiting for the ones who will take it – now, in its beginnings – as a sculptor takes his block of marble, and will shape it into a new beauty and a new reality'. But the adoption of television by a mass audience would have to wait until a post-war world was ready.

Stephen Herbert
Hastings, 2004

Part 5

A NEW ERA

High-definition and regular broadcasting in the United States of America

TELEFILMED FACES

Roosevelt and Landon Seen in Television Newsreel Test Across New York

By ORRIN E. DUNLAP Jr.

PRESIDENT ROOSEVELT and Governor Landon were telecast over the New York area a few days ago on the late evening air.

The few observers who have television receivers in their homes saw the two major nominees in the Presidential race talking to political gatherings as photographed by newsreel camera men. Mr. Roosevelt talked from the observation platform of a train. Mr. Landon addressed a crowd "somewhere in the East," according to those who looked in.

Pictures Clear on Long Island

The telefilm was run through the television transmitter atop the Empire State Building for broadcasting on tiny wave lengths. The "show" was so clear and the sound so loud on Long Island that it is believed the images were unreeled into space with sufficient power to overlap parts of at least three States, namely, New York, New Jersey and Connecticut.

A spectator described the quality of the images as excellent. In fact, he added that the reproduction was so "satisfactory" that there could be no doubt after seeing Mr. Roosevelt and Mr. Landon on the screen that political campaigning by television will be practical in the not-distant future.

The pictures were about six by seven inches in size and were screened without noticeable flicker.

The "show," which was on the air for more than an hour, also featured a film of the Spanish rebellion. The television "eye" was next turned on a large poster and several small objects, such as buttons on a dress, to demonstrate its ability in broadcasting all things which the electric camera "sees."

The pictures were telecast on the 49.75 megacycle channel and the associated sound on 52 megacycles. Both sight and sound leap from the same aerial rod simultaneously. The power of the ultra-short wave transmitter is rated at 10 kilowatts.

Improvements Are Expected

The pictures are broadcast at the speed of thirty a second and are comprised of 343 lines, the equivalent of the Electrical Musical Industries current television broadcasts from Alexandra Palace in London. The American engineers are understood to be developing a 441-line system, which will result in larger and clearer pictures.

The tests from the skyscraper have been under way since June 29. The engineers report that they have succeeded in "ironing the bugs" out of the transmitter and are now operating on a daily but irregular schedule.

The picture, as seen on the fourteen-control-knob set, has a greenish tint. The receiver is not as complicated to tune as the number of controls might indicate, for after the preliminary adjustments are made two or three knobs do the trick.

* * *

5

TELEVISION CALLED ANSWER TO SONG HOW TO KEEP 'EM DOWN ON THE FARM

TELEVISION is seen as a partial solution to keeping farm boys on the farm. T. Stewart Lyon, chairman of the Ontario Hydro-Electric Commission, told delegates to the Third World Power Conference that television might raise "a most formidable barrier to the drift of rural population to the cities."

Back in 1921 similar predictions were heard of radio spreading contentment on the rural acres. No doubt broadcast entertainment has banished many lonesome hours in isolated places. The radio statisticians estimate, however, that today only 2,000,000 of the 6,000,000 farms in the United States are radio-equipped. The extent of radio's influence in halting migration from farm to city is anybody's guess. Some say that when the farm boy and girl hear the city voices and plaintive melodies "the grass looks greener on the other side of the fence" and they are lured to the city.

There is the old song of World War days that asked the question "How are you gonna keep 'em down on the farm after they've seen Paree?" That may be more of a problem after they see by television not only Paree but other cities round the globe.

Still there are those who will argue if Ohioans see a telecast of Niagara, the Yosemite, or New York from atop a skyscraper, why go there? There may be a parallel with the case of radio and prize-fights. Once it was said people would not go to the ringside if they could sit at home and eavesdrop on the blow-by-blow description free. But that has not held true.

And so with television, it may merely advertise the world outside the farm, and while some may be content with the ethereal scenes, others will be stirred to go to the places they see on the television screens just as their curiosity is aroused to see Radio City and watch the studio show which they pluck from the air over the farm lands. The automobile makes it too easy to go to town to the movies. Television will be no substitute for travel; it will merely advertise what there is to be seen far across the orchards and the wheat fields.

Photos courtesy National Broadcasting Company

Scene in NBC experimental television studio in RCA building, New York, during a television fashion parade. Eighty-seven programs were telecast by NBC last year

TEN years ago a woman sat under blinding lights in John L. Baird's television studio in London while a group of men, assembled around a receiver in Hartsdale, N. Y., saw her face on a screen.

That radio transmission of a moving picture across 3,000 miles of ocean led many to believe that television, a new Twentieth-century wonder, was about to round the corner and, like radio, enter most American homes. But years passed and nothing of this sort happened. People still are asking, "When will we have television?"

There are three different answers to this question, all of them true. We have television right now—as a laboratory accomplishment. We have it also in the home—but it is limited to the homes of a few experimenters living within a few miles of a few stations. As far as most people are concerned, however, we do not have it at all. And no one knows, even now, when it will be ready for the general public.

Indeed, it might appear to the layman that television has moved backward, instead of forward, since 1928. Baird's historic telecast spanned an ocean, but the pictures which you may see eventually in your home probably will come from a transmitter, the effective range of which will be limited to the horizon line—twenty to fifty miles, depending on the height of the antenna.

If a 3,000-mile telecast was possible ten years ago, why are they limited to fifty miles now? There are several reasons.

Clearer pictures are being transmitted today and they are being transmitted over short waves, instead of long ones. If a wide band of long-wave frequencies were available, and if weather and other conditions are favorable, the range of good television signals is sufficient to transmit a low-definition picture clear around the globe, according to C. W. Farrier, NBC television coordinator.

But to transmit a picture of high definition by short wave and under varying con-

TELEVISION NOW ?

ditions, the dependable radius of a single transmitter has been found to be limited to the horizon—about as far as you can see. To send a program beyond the horizon without fear of failure, two or more transmitters must be connected by co-axial cable, the only known metallic conductor which can be used for telecasting.

And co-axial cable, capable of conducting fre-

Engineer at control board as he monitors television image. Above, focusing on Miss Patience in television studio before a telecast

quencies as high as 1,000 kilocycles or of carrying 200 telephone conversations at once, is expensive. But this is still inadequate for modern high-definition television transmission. One such circuit now connects New York and Philadelphia, but it will cost millions of dollars to crisscross the country with co-axial cables and link many transmitters together. Until this is done, however, there can be no chain television transmission like the radio broadcasts which go out to a network of stations from a central point.

Also, a television transmitter costs more to buy and to operate than a radio station of comparable quality. And even after

these costly transmitters are built and linked together by co-axial cable, only those living within a few miles of the transmitters will be able to receive pictures.

A television receiver is much more complicated than a sound receiver because it must be exactly in step with the transmitter to the millionth part of a second or there will be—no picture. Video, as television may soon be known, is a system of synchronizing and harmonizing many vital parts. If receivers and transmitters do not fit, as a key fits a lock, the system cannot function.

One stumbling block to public television today is lack of standardization. Uniform standards are more essential in television than in sound broadcasting because of the precise synchronization required between transmitter and receiver. Standardization will enable you to "look in" on two or three video programs in your territory instead of only one. Ten standards have been agreed on and a committee of the Radio Manufacturers' Association is attempting to agree upon others which will enable re-

white, "two" signifies gray and "three" black. By dividing the scene into imaginary squares, 100 each way, numbering them in sequence from one to 10,000, and sending a friend 10,000 digits, each representing the shade in one square, you would enable him to reproduce the scene on a sheet of paper ruled into 100 squares each way.

Obviously, this would be a tedious process. If, however, instead of sending numbers by telegraph code, you transmitted one

ceiver makers to provide instruments which will not become obsolete over night.

Despite the rather gloomy outlook for television for the masses in the immediate future, notable advances have been made in the art in recent years. One of the most important was Dr. Vladimir Zworykin's invention of electronic scanning which helped overcome the time element, one of the big problems in television, but of no consequence in sound broadcasting.

In sound broadcasting, Arthur Van Dyck of RCA points out, only one sound is transmitted at a time and this sound, even if it is a complex one, can be represented by one electric current. But no picture can be represented by one current or one anything else because it is composed of many elements. If you look at a scene ten feet square from a distance where the eye can see objects one inch in diameter, there are nearly 15,000 one-inch areas, which you must describe to convey the exact scene to someone else.

Given unlimited time, you might do this with a simple telegraph code. Suppose you arrange a code in which "one" means

Regulation television receiver and three-by-four-foot screen on which image was "stepped up." Above, "Kinescope" projection tube which cast enlarged image on screen

electric impulse for each square in sequence, the strength of each impulse corresponding to the degree of light in the square it represents, you might send a description of the picture in 10,000 seconds, or two and one-half hours, if you transmitted one impulse per second.

At the receiving end a printing device would be necessary to record each impulse in the same order and location and with an ink intensity corresponding to the current intensity of each impulse. The chief prob-

lem is one of synchronization between transmitter and receiver. This is the facsimile system, used on wire and radio, except that several impulses are sent each second, so only ten minutes or so are required to transmit a picture, instead of two and one-half hours.

But here's the rub. Television must transmit moving scenes, sending as many as thirty pictures per second so that, as in the movies, the eye will be deceived into believing it sees a continuous scene rather than a succession of stills. To send thirty pictures per second, transmitting information about each little part of each picture and repeating the process many times each second, a system 18,000 times faster than facsimile is required.

Here we have the primary cause of most of the television engineering problems—the time element, the necessity for transmitting an enormous amount of information very accurately and very quickly. Electronic scanning with the aid of the "Iconoscope" has helped solve the problem.

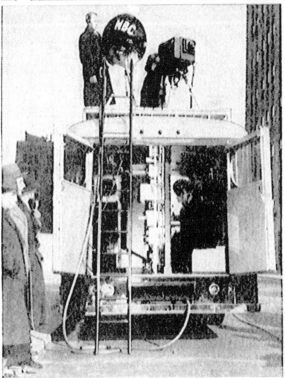

Mobile television unit which picks up both sight and sound. Top, girl as she appeared before "Iconoscope" camera and the 441-line televised image of girl as it appeared in television receiver

The Iconoscope converts light waves into electricity just as the microphone converts sound waves into electricity. The Iconoscope contains a plate upon which the scene being televised is focused. The surface of this plate is covered with thousands of photoelectric cells, microscopic in size, each separate from the others and each generating electric voltage proportional to the light which strikes it. To use these voltages, they must be collected from each cell.

This might be done by brushing a tiny wire across the plate, contacting the whole area bit by bit. But the idea is imprac-

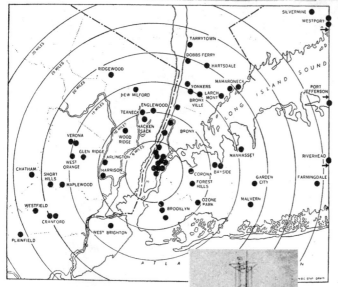

Photos at top © by Felix the Cat, Inc.

Above, left to right, image transmitted with sixty-line, 120-line and 441-line definition. Dots on map show locations of television receivers in New York metropolitan area which get programs from antenna atop Empire State building, below

441 lines is the best compromise between picture quality and apparatus difficulty and this number has been made standard for this country.

The beam explores the whole plate thirty times per second and there are about 250,000 spots to be thus visited. It "reads" from left to right at two miles per second, and from right to left at twenty miles per second. If your eye could move that fast, you could plow through a 1,000-page book in about five seconds. Thus this beam is about the busiest thing in this world.

The electric currents obtained by the beam from the cells are small but they can be amplified and then you have currents carrying intelligence representing the picture. These currents control the transmitter antenna current. At the receiving end, the entire process is reversed with the aid of the "Kinescope," the inverse of the "Iconoscope."

The Kinescope has a plate and beam of electrons playing

tical, partly because of the speed with which the operation must be performed. So a beam of electrons is utilized instead. This tiny "searchlight" travels over the plate, line by line, collecting the electric charge from each cell in turn.

The present standard calls for 441 of these lines on the plate from top to bottom, whereas the first systems had only twenty-four. The greater the number of lines, of course, the greater the detail of the pictures. But the more lines there are to cover, the more work the little beam must do and the more information there is to transmit. It has been found that upon it, just as does the Iconoscope. In the former, however, the plate or screen comprises one end of the tube itself, is made nearly flat, and coated on the inside with a thin layer of material which fluoresces, or gives off light, when electrons strike it. When the electron beam in the Kinescope strikes the screen in one spot, about the size of a pinhead, this spot glows, its brightness varying as the strength of the beam varies.

That spot of light is used to reproduce each spot of the picture, one at a time. The beam "paints" the lights and shadows of each tiny element of the picture as a series of spots, but to our slowly reacting eyes, the spots are not visible and the screen appears to be illuminated evenly all over. The flying beam of the Kinescope must be in perfect step with the flying beam of the Iconoscope, miles distant, or there will be no picture. Synchronization is one of the television problems which has been solved.

Scenes of any size can be televised by the transmitter but at the receiving end, the size of the picture is determined definitely by the size and brilliancy of the Kinescope screen. At present, there are two standard sizes, one about five by seven inches, the other about seven by ten inches. Last fall, NBC showed television on a seven-by-ten-foot screen, but there is a size limit for the tubes beyond which it is impractical to go. It has been found that the most desirable size of picture for television or movies is one where the height of the picture represents one-fourth the distance between screen and observer. In the home, the desirable viewing distance may be eight or ten feet, so the picture height should be about two feet.

"There is good promise of eventual accomplishment of this goal," says Mr. Van Dyck, "but at present it seems probable that the television receiver which is 'just around the corner' will have a picture about seven by ten inches."

Partly because so much information must be crowded into so small a period of time, tremendously high frequencies are necessary in television. Your light circuit

(Continued to page 139A)

RCA's Development of Television

At our annual meeting last year I announced that the Board of Directors had approved a plan for a field test in television. At that time it was stated that in twelve to fifteen months the project would be taken from the laboratory and subjected to field tests. I am pleased to inform you that our construction work has proceeded on schedule with the building of a new television transmitter located on the Empire State Building in New York, and with an experimental television studio in the RCA Building as a part of the National Broadcasting Company's operations.

Tests will start Monday, June 29, 1936. A number of experimental receivers are to be placed at observation points in the hands of our technical personnel so that we may determine the requirements and further development necessary to the establishment of a public television service.

This Corporation is second to none in the scientific and technical development of television. We have gone much beyond the standards fixed elsewhere for experimental equipment. But this is a far cry from the expectations of such a service aroused by pure speculation on the subject. There is a long and difficult road ahead for those who would pioneer in the development and establishment of a public television service.

RADIO CORPORATION OF AMERICA
RCA BUILDING
30 ROCKEFELLER PLAZA
NEW YORK

May 7, 1935

Release— IMMEDIATE

STATEMENT ON TELEVISION
By
David Sarnoff, President, Radio Corporation of America

At the Annual Meeting of RCA Stockholders
New York City, May 7, 1935

Public interest in television continues unabated since
the statement made in the annual report to the company's stockholders
on February 27, 1935. In that report it was stated that the management of the Radio Corporation of America was diligently exploring
the possibilities of a field demonstration, the next practicable step
in the development of television, in order that subsequent plans may
be founded on experience thus obtained.

As further stated in that report, our laboratory efforts
have been guided by the principle that the commercial application of
such a service could be achieved only through a system of high-
definition television which would make the images of objects transmitted, clearly recognizable to observers. The results attained by
RCA in laboratory experiments go beyond the standards accepted for
the inauguration of experimental television service in Europe. We
believe we are further advanced scientifically in this field than
any other country in the world.

In view of our own progress, and recent public statements on television made both here and abroad, I feel that the annual meeting of our stockholders is an appropriate occasion for a statement of Radio Corporation's position in television and of its plans for the immediate future.

First, let me emphasize that television bears no relation to the present system of sound broadcasting, which provides a continuous source of audible entertainment to the home. While television promises to supplement the present service of broadcasting by adding sight to sound, it will not supplant nor diminish the importance and usefulness of broadcasting by sound.

In the sense that the laboratory has supplied us with the basic means of lifting the curtain of space from scenes and activities at a distance, it may be said that television is here. But as a system of sight transmission and reception, comparable in coverage and service to the present' nation-wide system of sound broadcasting, television is not here, nor around the corner. The all important step that must now be taken is to bring the research results of the scientists and engineers out of the laboratory and into the field.

Television service requires the creation of a system, not merely the commercial development of apparatus. The Radio Corporation of America with its coordinated units engaged in related phases of radio communication services is outstandingly equipped to supply the experience, research and technique for the pioneering work which is necessary for the ultimate creation of a complete television system. Because of the technical and commercial problems which the art faces, this system must be built in progressive and evolutionary stages.

Considering these factors and the progress already made by your company, the management of the RCA has formulated and adopted the following three point plan:

THE PLAN

1. Establish the first modern television transmitting station in the United States, incorporating the highest standards of the art. This station will be located in a suitable center of population, with due thought to its proximity to RCA's research laboratories, manufacturing facilities, and its broadcasting center in Radio City.

2. Manufacture a limited number of television receiving sets. These will be placed at strategic points of observation in order that the RCA television system may be tested, modified and improved under actual service conditions.

3. Develop an experimental program service with the necessary studio technique to determine the most acceptable form of television programs.

Through this three point plan of field demonstration we shall seek to determine from the practical experience thus obtained, the technical and program requirements of a regular television service for the home.

It will take from twelve to fifteen months to build and erect the experimental television transmitter, to manufacture the observation receivers and to commence the transmission of test programs.

The estimated cost to the RCA of this project will be approximately one million dollars.

In order that the promise, as well as the present limitations, of the art be thoroughly understood, I shall review briefly the present status of television and the position of RCA in this field.

RECEPTION

Our research and technical progress may be judged by the fact that upon a laboratory basis we have produced a 343-line picture, as against the crude 30-line television picture of several years ago. The picture frequency of the earlier system was about 12 per second. This has now been raised to the equivalent of 60 per second. These advances enable the reception, over limited distances, of relatively clear images whose size has been increased without loss of definition.

From the practical standpoint, the character of service possible in the present status of the art, is somewhat comparable in its limitations to what one sees of a parade from the window of an office building, or of a world series baseball game from a nearby roof, or of a championship prize fight from the outermost seats of a great arena.

TRANSMISSION

In the present state of the art, the service range of television from any single station is limited to a radius of from fifteen to twenty-five miles. National coverage of the more than three million square miles in the United States would require a multitude of stations with huge expenditures, and presents a great technical problem of interconnection in order to create a network system by which the same program might serve a large territory.

Existing and available wire systems are not suitable for interconnecting television stations. Therefore, radio relays must be further developed or a new wire system created to do the job now being done by the wires which connect present day broadcasting stations.

While the problem of interconnection still remains one of the great obstacles to the extension of sight transmission and reception, an outstanding accomplishment in the field of television research is the invention and perfection by RCA engineers of the "iconoscope". This is an electric eye, which has advanced the technique of television by facilitating the pickup of studio action and permitting the broadcast of remote scenes, thereby giving to the television transmitter the function of a camera lens. Through the use of the iconoscope, street scenes and studio performances have been experimentally transmitted and received.

RELATIONSHIP OF TRANSMITTER AND RECEIVER IN TELEVISION

Television is a highly complicated system of transmitting and receiving elements with thousands of interlocking parts, each of which must not only function correctly within its own sphere of activity, but must also synchronize with every other part of the system. In broadcasting of sight, transmitter and receiver must fit as lock and key.

On the other hand, broadcasting of sound permits a large variety of receiving devices to work acceptably with any standard transmitter. Notwithstanding the great progress that has been made in sound broadcast transmission, a receiving set made ten years ago can still be used, although with great sacrifice in quality. This is not true in television, in which every major improvement in the art would render the receiver inoperative unless equivalent changes were made in both transmitters and receivers.

Important as it is from the standpoint of public policy to develop a system of television communication whereby a single event, program or pronouncement of national interest may be broadcast by sight and sound to the country as a whole, premature standardization would freeze the art. It would prevent the free play of technical development and retard the day when television could become a member in full standing of the radio family. Clearly, the first stage of television is field demonstration by which the basis may be set for technical standards.

ESTABLISHMENT OF TELEVISION AS A PUBLIC SERVICE

From the foregoing it will be seen that a number of basic problems surround any effort to establish television on a regular basis of public service to the nation. The more important of these problems are:

1. The fact that if the new art of television is to make the required technical progress, there will be rapid obsolescence of both television transmitters and television receivers.

2. The creation of new radio or wire facilities of interconnection before a service on a national basis can be rendered.

3. Further development, through experimentation in the field, of a system of high definition television which calls for new radio technique inside and outside the studio, and for the production of home television receivers which will increase the size of the picture and at the same time decrease the price at which the receiver can be sold to the public.

As in other related fields, the Radio Corporation of America is undertaking to encourage, develop and coordinate the research, engineering and technical processes by which a new art and a new industry may spring from the original root of radio communications. Television demands the most effective coordination of the equipment, facilities and services embraced by the operations of the various units of the RCA. It requires the utilization of the best engineering and manufacturing experience of the RCA Manufacturing Company in the production of television equipment; of the research facilities of the RCA laboratories for the development of new television tubes; of our experience in the construction of transmitting stations; of the studio, program, and broadcasting technique created by the National Broadcasting Company; and of the general experience of R.C.A. Communications.

Scientific research is the foundation of the radio industry and the achievements of RCA laboratories are most valuable assets.

These research activities have helped to maintain the Radio Corporation's position of leadership in the field of radio.

In announcing this plan, I wish to emphasize the clear distinction that must be made between the coming field demonstration stages of television, and the ultimate fulfillment of the promise of world-wide transmission of sight through space -- an achievement which will be second only to the world-wide transmission of sound through space. The sense of sight which television must eventually add to the body of radio communications cannot supplant the service of speech and music which permits any single event to be simultaneously broadcast to the nation as a whole, and which brings to millions of homes continuous programs of entertainment, information and education.

While the magnitude and nature of the problems of television call for prudence, they also call for courage and initiative without which a new art cannot be created or a new industry established. Your Corporation has faith in the progress which is being made by its scientists and its engineers, and the management of the Radio Corporation of America is exploring every path that may lead to an increasing business for the radio industry and to a new and useful service to the public.

- - - - - - - -

THE MESSAGE OF RADIO

Address by

David Sarnoff, President
Radio Corporation of America

--

Delivered June 29, 1936, at Hotel Statler, Boston, before the
Thirty Second Annual Convention of the Advertising Federation
of America.

--

We want to give him a seat in the front row
of the orchestra. When television broadcasting reaches the stage of
commercial service, advertising will have a new medium, perhaps the
most effective ever put at its command. It will bring a new
challenge to advertising ingenuity and a stimulus to advertising
talent.

The new medium will not supplant nor detract from the
importance of present-day broadcasting. Rather, it will supplement
this older medium of sound and add a new force to the advertiser's
armament of salesmanship. Television will add little to the
enjoyment of the symphony concert as it now comes by radio to your
living room. Sound broadcasting will remain the basic service for
the programs particularly adapted to its purposes. On the other
hand, television will bring into the home much visual material -
news events, drama, paintings, personalities - which sound can
bring only partially or not at all.

The benefits which have resulted from the industrial
sponsorship of sound broadcasting indicate that major television
programs will come from the same source. It requires little
imagination to see the advertising opportunities of television.
Broadcasting an actual likeness of a product, the visual

demonstration of its uses, the added effectiveness of sight to
sound in carrying messages to the human mind - these are only a few
of the obvious applications of television to merchandising. Commercial announcement can be expanded through television to include
demonstration and informational services that will be of value
to the public as well as to the advertiser.

Broadcasting has won its high place in the United States
because - unlike European listeners - American set owners receive
their broadcasting services free. Despite the greater cost of
television programs, I believe that owners of television receivers
in the United States will not be required to pay a fee for television programs. That is an aspect of the television problem in
which the advertising fraternity will doubtless cooperate in
finding the commercial solution.

Whoever the sponsor may be, or whatever his interests
or purposes, he will be under the compulsion to provide programs
that will bring pleasure, enlightenment and service to the American
public. That compulsion operates today and must continue to operate
if we are to retain the American system of radio broadcasting. The
public, through its inalienable right to shut off the receiver or
to turn the dial to another program will continue to make the rules.
In television as in sound broadcasting the owner of a set will
always be able to shut it off. In other words, the ultimate censorship of television, as well as of sound broadcasting, will remain
between the thumb and forefinger of the individual American.

COMMERCIAL TELEVISION — AND ITS NEEDS

By

DR. ALFRED N. GOLDSMITH
Consulting Industrial Engineer, New York

IT HAS been well said that anyone may be excused for making some mistakes, at least once — but that only a fool continues to make the same mistakes in the face of past experience. We in the radio industry cannot assert that we have made no mistakes in the past; despite the fine growth of the industry and its genuine contribution to our national life, there has been much friction and lost motion. And we now face a complicated situation in the case of television broadcasting which seems to impend. Shall we plunge wildly forward, substituting enthusiasm for analysis? Or shall we remain calm in our planning and activities, leaving the excitement to the properly persuasive advertisements of television receivers and programs, to the active salesmen, and to the delighted lookers and listeners in the home? With the confused past of radio in mind, shall we "try everything"—not once, but twice? Of shall we do as little "muddling through" as is humanly possible?

Commercial television broadcasting, to win general public acceptance and to enjoy a healthy growth, must be built on the basis of a group of necessary elements. These are a constructive Governmental attitude implemented by corresponding regulation, an active group of television broadcasting stations at least partly interconnected into national networks for program syndication, forward-looking program-building organizations, careful engineering and manufacturing methods rendered effective by suitable merchandising practices and satisfactory servicing, an enthusiastic and numerous group of home lookers, and finally a number of broadcast advertisers willing and able to secure the part-time attention of the home audience. To just the extent that any of these elements are missing from the television picture, the day of widespread public acceptance of television-telephone broadcasting will be delayed and the success of the radio industry reduced.

We believe that optimism for the future of television is well-founded, but we do not believe that the full measure of its success is attainable automatically and without at least some careful planning. Looking back on some of the shortcomings and mistakes of the past, we venture to hope that the future will bring the avoidance of at least some familiar and unnecessary errors. Accordingly it seems in order to study in more detail each of the elements of television success just mentioned, and to sketch the general outlines of the "paths to glory".

At the present time, the policy of the Government of the United States toward the development of television broadcasting is expressed through the radio laws passed by Congress and the regulations promulgated and decisions made under the authority of these laws by the Federal Communications Commission (and, in some cases, by affirming or reversing courts of appeal). It is stimulating to note that the Commission is evidently considering the needs of television broadcasting in an orderly and serious manner, and it is to be hoped that the decisions of the Commission in this field will be both generous and firm. To encourage the development of a new national industry is surely in accord with the spirit and needs of the times. One of the problems under consideration is that of television standards. It is neither easy nor economic, under known methods, to change television receivers from adaptability to a certain transmission to adaptability to a different transmission standard. In this respect television differs markedly—and unfortunately— from telephony; and this factor cannot be neglected in planning for television acceptance. It becomes necessary, from the very commercial beginning, to establish standards which have every likelihood of being satisfactory to the public for a long period of years in order to avoid speedy obsolescence of these early television receivers in the higher price ranges with consequent general dissatisfaction and loss of confidence. This is one case where we must "aim high" regardless of temptation to "cut corners". As we have repeatedly pointed out, the criterion of any television service is its continuing entertainment value; and it is now the consensus of more informed opinion that this specification requires pictures having of the order of 400 lines or more. Particularly is this the case if the pictures are to be increased in size from their present modest dimensions; and we do not doubt that such an increase will in due course be found commercially feasible. Another group of standards, in addition to the basic requirements of band width and ultra-high-frequency allocations, deals with picture repetition rate, scanning method, aspect of ratio, and synchronizing methods. Here again, the regulatory authorities will be well advised to require any essential uniformity even at the cost of some inconvenience to individual groups and in the interests of the general public.

In carrying out its tasks, the Commission will face the problem of allocating individual frequency bands for television to particular organizations in certain localities. For example, the question may arise: to whom shall be assigned, upon application, the available ultra-high-frequency bands in a given city for use in television broadcasting? There may well be numerous claimants actuated by a wide variety of motives. Some will desire to carry out a scientific experiment; others a commercial venture. Some will be highly experienced in broadcasting; others may be ambitious newcomers seeking a meteoric career in the television field. Some will have extensive technical and program background, while others will lack interest in the engineering and entertainment aspects of station operation. Much wisdom and restraint will be required of any regulatory body which faces the necessity for such decisions as are implied in the foregoing. We may venture the general suggestions that television broadcasting allocations should be granted to those who are best qualified by parallel experience, by technical and other resources, and who are most likely in the long run to keep abreast of engineering and program progress and the most modern operating methods. It is not our belief that, on the average, television broadcasting could be better handled by those who have had no previous experience with telephone broadcasting than by those who have carried the burdens and enjoyed the privileges of our present broadcasting system for many years. Nor do we see any compelling considerations in favor of granting television broadcasting priorities to expanding groups in the entertainment, news-disseminating, or advertising fields. It is more reasonable to expect that the healthy development of television and the solution of its highly specialised problems will result from an independent broadcasting industry rather than from a group of by-product activities of other (and seemingly competitive) industries. Accordingly it is urged that, in a reasonable time, there be granted the necessary allocations to qualified applicants for local transmitting rights in the commercial television broadcasting field.

In developing television broadcasting, it is necessary dispassionately to consider the best way of reaching a multitude of homes with program material of continuing interest. On one side of the ledger—the expense side—we find the cost of the transmitting facilities and of their operation together with the cost of creating the programs and syndicating them. It is a truism that the more persons reached effectively by a given program of quality, the clearer the justification for that program and its cost and the greater the likelihood that there will be a continuance of programs of like quality. We face then the dominant factor of the *program cost per listener*. This is the Sphinx at which every broadcaster thoughtfully stares, awaiting an answer to his questions. When it is considered that qualified artists, authors, arrangers, and directors are relatively shy and rare birds found in few localities (and therefore purchaseable only at a price) but that the audience is widely scattered, it again becomes evident that the only known way of reducing program-delivery cost per listener (which is another way of saying: increasing program quality per listener) is by program syndication. We are not here discussing the relative merits of various methods of total or partial syndication of programs such as transcriptions on wax or film, circulation of performers by road-shows, wireline or coaxial-cable interconnection of stations, and connection by radio relay systems. Nor yet are we considering the commercial and administrative aspects and problems of net work operation. We wish only to emphasize that syndication is of the essence of high-quality and stable television broadcasting and that it merits aid and support from all who are genuinely interested in the commercial success of that art.

Another nation-wide problem is that of avoidable man-made interference with radio reception. Like the poor, we have always had this electrical enigma facing us. An excellent beginning has been made in tackling this problem by the recently organized Sectional Committee on Radio-Electrical Coordination of the American Standards Association. But television broadcasting will be radiated in an unusually vulnerable region, namely the ultra-high frequencies. Automobile ignition systems and similar sources can superimpose on the television picture the perpetual twinkling of a myriad of stars—an effect as startling as it is unwelcome. There is a trend in some quarters to suggest legislation and resulting Commission action in the abatement of this trouble. One need not accept nor reject this suggestion in making the statement that, in one way or another, interference with satisfactory television reception at reasonable signal levels and for proper home installations *must* be accomplished. The prophecy can also safely be made that it *will* be accomplished since millions of otherwise embattled lookers (who happen also to be voters) will receive friendly consideration by prudent powers that be.

It is fortunate for television development that one of the ancient and popular fallacies of broadcasting—namely, the opposition to high-power, so called, in transmission—has largely had its day and been relegated in the main to the dust pile of forgotten errors. We can well remember the learned gentleman who, little more than a decade ago, hotly informed a gaping radio conference that no one needed a higher transmitting power that the half-kilowatt of his own station; and who justified this by stating that he had "national coverage" as evidenced in the form of letters from most states of the Union where listeners had heard his station! Were this gentleman dead—which we are happy to say he is not—he would undoubtedly turn in his grave at the solemn proposal to limit the power of a certain class of broadcasting stations to *not less than 50 kilowatts* with the hope that 500 kilowatts will be widely used. Thus we may expect that powers of tens of kilowatts or more for television stations in the larger cities will be taken as a matter of course and wisely regarded as the boon to good service which such stations actually are. In this regard, at least, the Commission inherits a well-ploughed and fertile field.

Any reviewer of television needs would be but too happy to find no occasion to mention such matters as the briefness of assured tenure of television licenses, on the one hand, and the apparent occasional interaction of governmental action and political considerations on the other hand. Yet the picture would not be complete were these factors to be omitted nor would justice to the future television audience be done. We believe that few men would be willing to build a home or erect a factory if their leasehold were limited to uncertain successive six-month periods. Particularly would this be the case if there were much of pioneering and risk in their building activities. Those who bravely enter into television development in these parlous times are entitled to friendly consideration and no little encouragement. We submit the thought that it will be a stimulant to television development to relieve those who enter it from the parallel and non-profitable activity of prosecuting applications for license renewals in a steady stream and to permit them to bring a peaceful and unapprehensive mind to bear on their major (and sufficiently difficult) problems of providing good television service. A considerable extension of the license period for broadcasting would be a genuine contribution to the development of that field.

Insofar as broad questions of the public welfare are involved in legislation, regulation, and external administration of radio broadcasting, the influence of the statesman and legislator is to be sought. But whenever questions of complicated technical nature, of internal administration, and of detailed planning are before the Commission or the broadcasters, we have little expectation of real assistance from politicians. Broadcasting has been an effective means for bringing the views of the political leaders of this country to their constituencies. Enlightened self-interest will prompt a corresponding policy of non-interference with television development on their part or disregard of attempted interference.

The second necessary element for television success is a group of well-constructed and capably managed transmitting stations with a suitable measure of interconnection for program syndication. It is thus incumbent on the present-day broadcasters and networks to take up the burden of establishing the necessary facilities. Only by so doing can they hope to assume that position in the future television set-up to which they appear normally entitled. That television will come can hardly be doubted. If it does not come through an expansion of the facilities of those now engaged in telephone broadcasting, it will come through the enterprise of others—and, as we have previously indicated, this is not in our opinion a desirable process of evolution so far as a healthy and normal growth of television is concerned.

It is necessary that there shall be available transmitters and studio equipment of suitable powers for various localities —and it is fair to assume that the manufacturers of such equipment will make it available as soon as their own tests of sample models shall have inspired confidence in the performance of the available equipment. The engineers are now a part of a great movement in the direction of commercializing television, and any engineer who is giving his best efforts to the improvement of transmitting or receiving equipment of that type is taking his earned place abreast of the radio pioneers of the past. The manufacturing of transmitting equipment will also do well to remember that the initial impressions of the public will largely depend on the quality of the transmissions, and that the best that can be produced will probably be none too good. Let there be no casual or careless transmitter production for television, in the interest of the entire industry as well as the public.

We have noticed with some concern what is, in the last analysis, an occasional and largely meaningless friction between the so-called local stations and the networks. As well might the hand object to the arm. Networks and outlet stations are an organic unit, and each equally needs the other. Stations not connected to any network render a related and parallel service. Dissension in these quarters represents merely an army divided within itself and thus facing combat with less chance of winning. To some extent, every entertainment industry tries to secure its share of the attention and leisure time of the public. Those industries which overlook small jealousies or fancied advantages in the interests of general advance will make the best public showing; and this is always reflected in that grim index of public acceptance: the balance sheet. For this reason the growth of cordial relations within the broadcasting groups is greatly to be desired; and we feel fairly confident that common sense will triumph to that effect. Then, too, if television broadcasters desire due consideration to be paid by legislators to their views, they must not present the unsavoury spectacle of a house quarreling within itself and speaking only through an angry babel of conflicting voices. The broadcasting organizations should rise to this opportunity to speak for all and with approval from all, on a basis so broad and farseeing that they will compel the attention and win the acquiescence of reasonable men.

No one can long study television broadcasting without becoming somewhat concerned as to the mode of program-department organization and the subsequent production of the necessary program material. The need for progressive program-creating groups in television broadcasting will be great indeed. On any reasonable standard of appearance and performance, it is clear that there are not available in clamoring throngs the necessary regiments of satisfactory performing artists for the new field. The stage and screen have preëmpted (at substantial cost) those who are judged most worthy of winning public favor through their appearance and performance. In the radio field, performance only (and that in the restricted range of sound) has hitherto been the sole criterion of artist success. Now the requirements broaden—but the supply of available talent does not. There should be a host of opportunities for a new generation of artists having "television personalities"— and what is the basis of such personalities, only time and experience will disclose. The grist to the artists' mill may well be numbered in tens of thousands of candidates; the chaff in thousands; and the finely winnowed grain in hundreds. Yet only along this hard road can be found those who will successfully face the public on the television screen.

Radio telephone broadcasting evolved, as an entertainment enterprise, in a curiously segregated fashion. Its relationships with other branches of the entertainment field have mostly been conspicuous by their absence. The phonograph, the disc record, the legitimate stage, the motion picture studio and theater, musicians, authors, copyright owners, actors, instrumentalities and soloists have all been involved in radio development in one way or another, but he would be bold who would say that radio had made the most of its contacts with these fields or that the relationships which existed were, in the main, more than haphazard and occasionally unhappy. Television broadcasting faces similar, but even more complicated and trying situations in this regard, and it would be well to remember that disregard or hostility, together with non-cooperation, rarely accomplish much. This thought applies, of course, not only to the radio facet of the situation, but to all the other interests involved. We have particularly in mind that the motion picture field (which, in our opinion, has little to fear from television broadcasting if it maintains a forward-looking outlook and is well guided), will have methods aid output which can be somewhat adapted to the needs of a certain part of television procedure. Television can, in turn place at the disposal of the motion picture industry certain new methods and devices which should be useful. Certainly the relationship between these sister arts could and should be pleasant and mutually helpful, in the best interests of each.

At this point we urge that television broadcasting adopt the desirable practice of totally excluding the public from actual attendance at all rehearsals and broadcast events in the studio. It must be remembered that broadcasting aims to serve the millions of its radio audience in the home. It should not be so necessary to stress that it is not properly a mode of amusing the advertising sponsor (who should have other and more practical aims) nor yet of entertaining the client's advertising agency. There is no question that the program timing and the methods of production suffers when a compromise between the home audience and the studio audience is adopted.

It is also well known in the entertainment field that the illusion is spoiled when the audience "sees the wheels go round". What stage magician would show the audience how he performs his tricks? What great dramatic company on the stage would invite the audience to watch the scenery being shifted and to inspect the prompter at his work? Even the concert soloist keeps himself in desirable seclusion until the moment comes for him to step upon the stage. The motion picture industry has well realized this and has practically closed its studios to the public. This wise measure might well be adopted by television broadcasters who are in a similar position as regards entertainment possibilities.

There are other incidental advantages in the elimination of the studio audience. Annoyance resulting from alleged competition with theaters, and the municipal licensing problems which accompany theater operation would be eliminated. Economy in operation is effected and greater freedom in procedure without interference. The mixture of film sequences and personal performances is more readily developed. The locating of television studios is simplified when only the home audience need be considered, and the studio arrangements and facilities will be more efficient and cost considerably less. Then, too, the rapid switching from one studio to another or the introduction of long or short motion picture subjects into the program can be carried out without worrying about the supposed reaction on the visitors to the studio.

Speaking frankly, we realize that building studios for home audiences only, conducting them in a practical, modest, and businesslike way; and retaining the illusion and consequent enjoyment of the home audience will involve some sacrifice of vanity on the part of client, agency, and broadcaster alike. But, as has been said, the entertainer is a vendor of illusion and a seller of glamor—or else he is nothing. Why then should he deliberately destroy part of his stock in trade? We have often heard persons who have just left a studio broadcast protest that they would not enjoy radio nearly so much now that they had seen the way in which program matters were actually handled or had a clearer picture of their favorite and previously idealized performer. We have listened to the annoyed protest of those who conduct broadcasting and who are compelled to go through useless motions and elaborate procedure for the hundreds in the studio in disregard of the millions in the home. Let television broadcasting, at least, be democratic and devote its efficient, concentrated, and exclusive attention to the home audiences who purchase the receivers, who watch the performances and who give television broadcasting its very life.

So far as the commercial leaders and engineers of the radio manufacturing industry are concerned, their tasks in the new television field will be heavy indeed. Every mistake or omission of the past should be carefully remembered and as sedulously avoided in the future. If the industry elects to make a "bread-board model" of a television receiver one day and to turn out allegedly commercial manufactured product immediately thereafter, without adequate field tests and painstaking engineering study and improvement in the interim, the public will gain an unfavorable impression of the quality, performance, and returns of the resulting product. It is hardly possible to devote too much care to the engineering design and test of the first large group of television receivers which the public purchases. A negative first impression at this point will take years to eradicate. And in such engineering work, let us remember that, although the skilled technician can handle a multiplicity of new, complicated, and delicate adjustments, the average tired man or woman at home neither can nor will go to the trouble of learning how to juggle a small-scale switchboard nor expend the time and effort necessary to continue to use the electrical cross-word which is thus presented. In other words, the engineers must be "home-minded" and leave the rarefied air of high but complicated technical achievement to come down to the lower and safer levels of simplicity and comfort in the use of television receivers.

Television comes into the world at a time when its nearest relatives have grown to maturity. For a relatively small sum, the public can see large, clear, and well-planned sound-motion pictures in impressive surroundings and under favorable physical and psychological conditions of presentation. In the home, not all the conditions are so favorable. Noise, stray light, interruptions both natural and man-made, inadequate seating arrangements for the audience and the like must be anticipated. Thus, we need the brightest, sharpest, and largest picture which can be economically and technically produced; and we must continue to improve the picture (and sound) in these regards as time goes on to hold public favor.

Further, it would be a wise investment to enlarge markedly the testing and supervisory force in factories devoted to television receiver manufacture. At best these new devices must be expected to develop unexpected troubles; and it is far better to discover these in the comparative privacy of the factory test than in the glaring spotlight of public indignation. Then, too, it will not be enough for the receiver to work as it leaves the factory. Every part should be carefully studied to make sure that it will stand up. It must be remembered that skilled television service men will not be too plentiful for the first months or years of television commercialization; and a dark television screen is as unattractive to the purchaser as a silent loud speaker.

We may presume that television circuits and models will naturally change fairly rapidly during the years of the introduction of television on a large scale to the public. This being the case, it is inadvisable further to complicate the commercial situation by exaggerating this tendency through the deliberate introduction of inconsequential or even imaginary "improvements", so-called, in reasonably satisfactory receiver models. A firm hand will be required on the commercial helm in this regard in every television-receiver factory. We must not be misinterpreted as regarding the early introduction of an actual and marked improvement as undesirable; we mean rather than we urge that television manufacture be not made the "happy hunting ground" of mere "gadgeteers".

Since, as has been mentioned, the quality of the television image can be affected by man-made static, associations of radio manufacturers should give every possible aid and encouragement to any groups that aim to reduce such interference. Information on installations, interference-reducing devices and methods, and cooperation with the automotive industry and the makers of diathermal equipment are matters of concern to the manufacturers. To the extent that these are handled and funds for their prosecution made available, the advent of wide-spread television sales will be hastened.

The service problem for television should not be left to grow at random as it largely did until recently in the case of present-day broadcasting. Training of service men by the manufacturers, radio schools, and associations of service men are notably in order. Since it wil probably take a fair time to train a man to locate trouble in so elaborate a device as a television-telephone receiver and then to repair the fault, early consideration should be given to this need for training.

We have noticed without pleasure or approval some of the published material of a rather wild sort dealing with television. The implication of such sensational statements is that the fortunate owner of a cheap television receiver, seated in a comfortable armchair in his home, will touch a button and on the opposite wall will appear what looks like a huge motion picture in color, with sound, which reaches him by television. A twist of the tuning dial and he will see at will a battlefield abroad, a performance equal to the finest feature films, a football game, or whatever other delightful performance his fancy can conjure up. Without wishing in any way to present a gloomy picture of what will actually occur, it is fair to say that those who expect what has just been described will be disappointed by the actual performance. A reasonable restraint in all statements made by individuals or by associations of manufacturers, of broadcasters, and of engineers will be useful in enabling performance to realize or, still better, to exceed expectations.

This leads to the final thought that the leading associations in the radio field should continue to cooperate fully in the interest of the successful commercial development of television broadcasting. Whether the existing contacts are close enough and extensive enough might well be submitted to keen scrutiny in the light of the need for most effective industry action, to the end that the fine possibilities of television broadcasting may be fully realized to the satisfaction of the public and the industry alike.

TELEVISION AND THE ELECTRON

By

Dr. Vladimar K. Zworykin

Director, Electronic Research Laboratory, RCA Manufacturing Company, Inc.

A FEW months ago the R.M.A. decided upon a set of standards to be applied to commercial television receivers. Among the requirements specified were those that the system should produce a 441-line picture with a picture frequency equivalent to 30 per second. These standards were accepted by the television engineering world with the utmost complacency.

If such a set of standards had been announced a few years ago it would have been instantly branded as quixotic idealism by almost every worker in the field. When it is realized that television research has been actively carried on for the past quarter of a century, this rapid advance in the last few years takes on real significance.

The cause of this extremely rapid advance which has changed television from a laboratory plaything into a practical engineering accomplishment was primarily a change from mechanical methods of picture transmission and reception to cathode ray systems.

Pioneering work in the field of cathode ray television had been carried on by a few isolated workers for a number of years previous to its general recognition by the major research laboratories. The work of these men served to illustrate to the world that the basis of cathode ray television was sound, and that electronic methods offered a solution to such problems as those of obtaining sufficient illumination on the viewing screen, of inertialess scanning required to obtain high definiton, and sufficient sensitivity for the successful transmission of pictures under ordinary conditions of illumination.

Once the way had been pointed out, a number of the more farsighted of the television research laboratories initiated a program of intensive research along this line. This work has been going on for the past five years and has led not only to refining

Reprinted from *Short Wave and Television*, April, 1937.

the basic principles advanced by the pioneers but also to the discovery and adaptation of a great number of new principles. As a consequence of this effort, both the television transmitter and receiver have become a practical reality.

The television receiver as it is today—using the *Kinescope** —resembles, in appearance and size, a console radio receiver. The reproduced picture is sufficiently brilliant to watch without strain in a moderately lighted room and is in size about a page of this magazine. Thus, while such a reproducing device is a long way from ideal, it nevertheless is capable of bringing to the observer a picture that has high entertainment value, one which is both pleasing and informative.

The pickup camera employing the *Iconoscope** is but little larger than a commercial 35 mm. moving picture camera, and since it contains no moving parts can easily be made portable. At its present stage of development its sensitivity is sufficient to enable the transmission of an out-door scene under almost all conditions of lighting, or a studio picture when bright but not uncomfortable lighting is used.

The picture signal from this and accompanying sound pickup is carried to the main ultra-short wave transmitter through cable or radio relay, and from there it is transmitted on a carrier of 5 or 6 meter wavelength. Such a transmitter is capable of servicing a radius of from 30 to 50 miles, depending upon the topography of the terrain.

Of course, it must be recognized that the problems of covering the country with a network of television transmitters, of manufacturing a reasonably priced receiver, and those involved in organizing and producing suitable programs are enormous. These problems are ones that must and will be met by the manufacturer, the production engineer and the technician. This solution is only a matter of time.

Even if some inconceivable law should come into existence that prevented the application of any new principles or developments to the cathode ray television system as it stands today, I am convinced that it would still become a commercial reality, that the system is amply capable of producing a picture which would satisfy a real economic demand.

However, this is equivalent to saying that the automobile of 1910 was a commercial reality. Certainly it was a mode of transportation which met a definite demand, and if all development

---

* Registered Trademark of the RCA Manufacturing Co., Inc.

had ceased at that date the automobile would still be extensively used today. Just as the useful but crude vehicle of 1910 has evolved into the luxurious motor car of today, which in its turn will be supplanted by an even better vehicle in the future, so the application of the laboratory research which is going on today must inevitably lead to improvements in the cathode ray television system.

Of course, the statement that marked advances in cathode ray television can be made is not proof that this progress is possible. However, research which is being carried out in the laboratory gives ample evidence of the improvements that may be expected as our knowledge increases. To give a concrete example, recent advances in electron optics makes it possible to produce an electron copy of a visible image and secondary emission, which has only just begun to be seriously studied, makes it possible to intensify this copy. These two new principles have been applied to laboratory models of the Iconoscope with a consequent many-fold increase in sensitivity. Another example that might be cited is that of the viewing tube. The size of the present television picture is limited because it is viewed directly on the fluorescent screen of the Kinescope and, consequently, is dependent on the physical size of the tube. *Laboratory experiments indicate that there is every reason to believe that it will be possible to build tubes giving a small picture of sufficient brilliancy to be projected upon a large viewing screen.* Experimental models of this type of projection tube have been made which very nearly meet the requirements of television. Continued improvements in the electron gun and in fluorescent material will unquestionably make this type of Kinescope entirely practical.

These are only two of the many examples that might be given of the progress that may be expected. Next year and the year after, examples which do not exist today can be given. In other words, the electron system has not yet even emerged from early childhood. Only the most incorrigible pessimist, the man who has an honest doubt about the sun's rising tomorrow, believes the cathode ray television is a closed field, that all is known about it that can be known.

Assuming that the system as it stands today can produce a fairly satisfactory picture and that there is every reason to look for marked improvements in the near future, lest us ask what will be required of television if it is to become popular in the sense that radio broadcast is popular.

Considering first the receiver, the entertainment supplied by the receiver must be such that it can be made incidental to the normal household activities. In other words, television is not and should not be intended to take the place of the observer's going in person to see an event in which he is intensely interested. The sport fan will still go to the baseball field, the football game or the boxing arena, the theatre lover will still go in person to see the plays in which he is interested, television or no television. However, to the individual who is not sufficiently interested in an event to expend the time and effort to become an eye witness, television will bring a summary of what is taking place. This means that the receiver must be small enough so that it will not be objectionable as a piece of furniture. It must be simple in operation and arranged so that it does not require setting up of viewing screens or any other elaborate preparation. The picture should be bright enough so that it can be readily seen in a moderately lighted room, and small enough not to be too obtrusive, perhaps one-and-a-half by two feet in size. In a sense, the receiver might be considered as a window through which the individual may, in the course of conversation or reading, glance to see what is going on in the world around him.

The television pickup device, to be completely satisfactory, must be sufficiently sensitive not only to reproduce scenes of average illumination but should also be operative at very low light levels. Imagine the feelings of the spectators looking at a football game if the last few minutes' play cannot be transmitted due to insufficient light. The Iconoscope of today, while it will suffice for ordinary weather conditions, would not be operative in the semi-darkness of late afternoon in November. However, as was pointed out above, there is every reason to expect a continuous improvement in the *sensitivity* of the Iconoscope as time goes by. Eventually, the Iconoscope may equal or even exceed the photographic camera in sensitivity.

Perhaps the most difficult to attain is a satisfactory *network of transmitters*. At present, the range of an individual transmitter is limited to the visual horizon as seen from its antenna. This means that the area serviced by a transmitter is relatively small, and that each urban center must have its own television transmitters. It is obviously necessary, in a completely satisfactory system, to be able to chain these transmitters in such a way that events can be broadcast nation-wide. These chains will be formed by inter-connecting the stations with means of *concentric cable* and by the use of *radio-relay links*.

This ideal system will eventually exist, but only after years of television broadcasting experience. In the meantime, we will have to be content with a much less perfect system. All the units for satisfactory television are ready and now await commercialization by those responsible for the economic and production aspects of the problem. But, as warning to those who are unduly optimistic, the problem of assembling these elements is almost as formidable as that of developing cathode ray television. Universal television in the home will not be an accomplished fact for a number of years to come but, on the other hand, it is absolutely assured that home reception of pictures will eventually be commonplace.

TELEVISION STUDIO TECHNIC*†

By

ALBERT W. PROTZMAN

Engineering Department, National Broadcasting Company, Inc.,
New York, N. Y.

Summary—The studio operating technic as practiced in the NBC television studios today is discussed and comparisons are made, where possible, to motion picture technic. Preliminary investigations conducted to derive a television operating technic revealed that both the theater and the motion picture could contribute certain practices.

The problems of lighting, scenic design, background projection, and make-up are discussed, with special emphasis on the difficulties and differences that make television studio practice unique.

An explanation is given of the functioning of a special circuit used in television sound pick-up to aid in the creation of the illusion of close-up and long-shot sound perspective without impracticable amount of microphone movement. The paper concludes with a typical television production routine showing the coördination and timing of personnel and equipment required in producing a television program.

IF ONE were forced to name the first requirement of television operating technic and found himself limited to a single word, that word would undoubtedly be "timing." Accurate timing of devices and split-second movements of cameras are the essentials of television operation. Personnel must function with rigid coördination. Mistakes are costly—they must not happen—there are no second chances.

Why such speed and coördination? Television catches action at the instant of its occurrence. Television does not allow us to shoot one scene today and another tomorrow, to view rushes or resort to the cutting room for editing. Everything must be done as a unit, correct and exact at the time of the "takes"—otherwise, there is no television show.

Now, to discuss some preliminary investigations conducted before production was attempted, and to describe the equipment and technic used in meeting these production requirements. Technical details are deliberately omitted. Wherever possible, we shall compare phases of television operation with their counterparts in motion picture production.

For so new a medium as television it is, of course, an impossibility to present a complete and permanently valid exposition. Television technic and apparatus constantly advance. Some technic now current may be outmoded in a day or a month. We have only to recall the early days of motion picture production, when slow-speed film and

---

* Decimal Classification: R583.2.

† Reprinted from the *Jour. Soc. Mot. Pic. Eng.*, July, 1939.

inferior lenses were a constant limitation. So, with television, it is already possible to envision more sensitive pick-up tubes that will permit the use of smaller lenses of much shorter focal length, thus eliminating many of today's operating difficulties.

PRODUCTION TECHNIC INVESTIGATIONS

In May, 1935, the Radio Corporation of. American released television from its research laboratories for actual field and studio tests. Long before the first program was produced in the middle of 1936, plans were laid, based on extensive research into the established entertainment fields, for the purpose of determining in advance what technics might be adaptable to the new medium of television. From the stage came the formula of continuity of action, an inherent basic requirement of television. This meant memorized lines and long rehearsals. Prompting could not be considered, for, as you know, the sensitive microphone which is as much present in television as it is in sound motion picture production, does not discriminate between dialog and prompting.

From the motion picture studio came many ideas and technics. If television is a combination of pictures with sound, and it is, no matter what viewpoint is taken, the result spells in part and for many types of programs, a motion picture technic at the production end. However, enough has already been said about the peculiarities of television presentation to justify saying that the movie technics do not supply the final answer. There remained the major problem of preserving program continuity without losing too much of motion picture production's flexibility. Our present technic allows no time for adjustments or retakes. Any mistake immediately becomes the property of the audience. The result of the entire investigation led to what we think is at least a partial answer to the problem. This technic, we hope, will assist considerably in bringing television out of the experimental laboratory and into the field of home education and entertainment.

GENERAL LAYOUT OF FACILITIES

In order to present a clearer view of our problems, we shall give a brief description of our operating plant. The present television installation at the National Broadcasting Company's headquarters in the RCA Building, New York, N. Y., consists of three studios, a technical laboratory, machine and carpenter shops, and a scenic paint shop. Our transmitter is located on the 85th floor of the Empire State Building. The antenna system for both sight and sound is about 1300 feet above the street level. Both the picture and sound signals are relayed from the Radio City Studios to the video and sound transmitters either by coaxial cable or over a special radio link transmitter.

One of the studios is devoted exclusively to televising motion picture film, another to programs involving live talent, and the third for special effects. It is the operation of the live-talent studio with which we are concerned in this paper.

Fig. 1(a)—General layout of live-talent studio; control room at upper rear.

DESCRIPTION OF LIVE-TALENT STUDIO

Figure 1(a) shows the general layout of the live-talent studio. The studio is 30 feet wide, 50 feet long, and 18 feet high. Such a size should not be considered a recommendation as to the desired size and proportions of a television studio. The studio was formerly a regular radio broadcasting studio, not especially designed for television. To anyone familiar with the large sound stages on the motion picture lots, this size may seem small (Figure 1(b)). Yet, in spite of our limited space, some involved multi-set pick-ups have been successfully achieved by careful planning. Sets, or scenes, are usually placed at one end of the studio. Control facilities are located at the opposite end in an elevated booth, affording full view of the studio for the control room staff. Any small sets supplementing the main set are placed along the side walls as near the main set as possible, and in such position as to minimize camera movement. At all times, we reserve as much of the floor space as possible for camera operations and such floor lights as are absolutely essential. At the base of the walls and also on the ceiling are scattered numerous light-power outlets to minimize the length of lighting cables. At the rear of the studio is a permanent projection room for background projection.

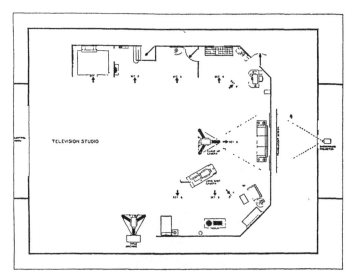

Fig. 1(*b*)—Television studio floor plan.

CAMERA EQUIPMENT

The studio is at present fitted for three cameras. To each camera is connected a cable. This cable is about two inches in diameter and fifty feet long; it contains 32 conductors including the well known coaxial cable over which the video signal is transmitted to the camera's associated equipment in the control room. The remainder of the conductors carry the necessary scanning voltages and current supplies for the camera amplifiers, interphone system, signal lights, etc. From this description, it is apparent that adding another camera in a television studio involves a much greater problem than that of moving an extra camera into a motion picture studio. In television, it is necessary to add an extra rack of equipment in the control room for each additional camera.

MOVEMENT OF CAMERAS

One camera, usually the long-shot camera using a short-focal length lens, is mounted on a regular motion picture type dolly to insure stable movements. The handling of the dolly is done by a technician assisting the camera operator. It is impracticable to lay tracks for dolly shots as is often the motion picture practice, because usually each camera must be moved frequently in all directions during the televising of a studio show. Naturally, dolly tracks would limit such movement. The other television cameras utilize a specially de-

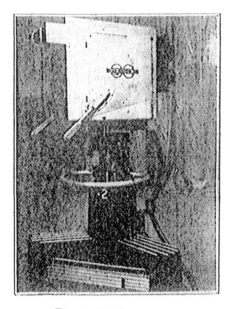

Fig. 2—Studio camera.

signed mobile pedestal (Figure 2). Cameras mounted on these pedestals are very flexible and may be moved in and out of position by the camera operators themselves. Built into the pedestals are motors which elevate or lower the camera; this action is controlled with push-buttons by the camera operators. A panning head, similar to those used for motion picture cameras, is also a part of the pedestal. It is perhaps needless to stress here that one of the strict requirements of a television camera is that it must be silent in operation. In the electronic camera proper there are no moving parts other than those used for focusing adjustments; hence, it is a negligible source of noise. When camera pedestals were first used they were the source of both mechanical noise and electrical disturbance when the camera-elevating motor was in use. Since then this problem has been overcome, and it can be stated that the entire camera unit is now free of objectionable mechanical noise or electrical surges.

Lens Complement

Each camera is equipped with an assembly of two identical lenses displaced 6 inches vertically. The upper lens focuses the image of the scene on a ground-glass which is viewed by the camera operator. The lower lens focuses the image on the "mosaic," the Iconoscope's light-sensitive plate. This plate has for its movie counterpart the film in a motion picture camera. The lens housings are demountable and interchangeable. Lenses with focal lengths from 6½ to 18 inches are used at present. Lenses of shorter focal length or wider angle of pick-up can not be used since the distance between the mosaic and the glass envelope of the Iconoscope is approximately 6 inches. Lens changes can not be effected as fast as' on a motion picture camera, since a turret arrangement for the lenses is mechanically impracticable at present. However, it is probably safe to say that future advances in camera and Iconoscope design will incorporate some type of lens turret. Ordinarily, one camera utilizes a 6½-inch focal length lens with a 36-degree angle, for long shots, while the others use lenses of longer focal lengths for close-up shots. Due to its large aperture, the optical system used at present has considerably less depth of focus than those used in motion pictures, making it essential for camera operators to follow focus continuously and with the greatest care. This limitation will probably be of short duration, since more sensitive Iconoscopes will permit the use of optical systems of far greater depths of focus.

It is desirable here to point out a difference in focusing technic between motion picture cameras and television cameras. "Follow-focus" in motion pictures occurs practically only in making dolly shots. For all fixed shots, the lens focus is set, the depth of focus being sufficient to carry the action. Also, it is the duty of the assistant cameraman to do the focusing. This relieves the cameraman of that responsibility and allows him to concentrate on composition, action, and lighting. In television, the camera operator must do the focusing for fixed shots and dolly shots alike. This added operation, at times, is quite fatiguing.

Vertical parallax between the view finder lens and the Iconoscope lens is compensated for by a specially designed framing device at the ground-glass that works automatically in conjunction with the lens-focusing control. It may be of interest to note here that at first the television camera had no framing device. This meant that images, in addition to being inverted as they are in an ordinary view-finder, were also out of frame. The camera operator had to use his judgment in correcting the parallax. With this new framing device, the operator now knows exactly the composition of the picture being focused on the mosaic in his camera. The framing device can be quickly adjusted to accommodate any lens between 6½ and 18 inches focal length.

Because of the fact that several cameras are often trained on the same scene from various angles, and because all cameras are silent in operation, performers must be informed sometimes—such as when they are speaking directly to the television audience—which camera is active at the moment. Two large green bull's-eye signal-lamps

Fig. 3—Typical television set.

mounted below the lens assembly are lighted when the particular camera is switched "on the air."

SET LIGHTING

There are two outstanding differences between television lighting and motion picture lighting. A much greater amount of key light is required in television than in motion pictures. Also, a television set must be lighted in such a way that all the camera angles are anticipated and properly lighted at one time. Floor light is held to a minimum to conserve space in assuring maximum flexibility and speed of camera movements. Great care must also be taken to shield stray light from all camera lenses. This task is not always easy, since, during a half-hour performance, each camera may make as many as twenty different shots. Just as excessive leak-light striking the lens will ruin motion picture film, it has a definitely injurious effect upon the photosensitive mosaic and upon the electrical characteristics of the Iconoscope.. A direct beam of high-intensity light may temporarily paralyze a tube, thus rendering it useless for the moment.

SETS

Television sets (Figure 3) are usually painted in shades of gray. Since television reproduction is in black and white, color in sets is

Fig. 4—Background projection window shot.

relatively unimportant. Chalky whites are generally avoided because it is not always possible to keep "hot lights" from these highly reflective surfaces which cause a "bloom" in the picture. This, in turn, limits the contrast range of the system. Due to the fact that the resolution of the all-electronic system is quite high, television sets must be rendered in considerable detail, much more, in fact, than for a corresponding stage production. As in motion picture production, general construction must be as real and genuine as possible; a marked difference, for instance, can be detected between a painted door and a real door. On the legitimate stage, a canvas door may be painted with fixed highlights; that is, a fixed perspective, because the lighting remains practically constant, and the viewing angle is approximately the same from any point in the audience. But, in television the perspective changes from one camera shot to another. Painted perspectives would therefore be out of harmony with a realistic appearance. This is also true in motion picture work. Sets must also be designed so that they can be struck quickly with a minimum effort and noise because it is often necessary to change scenes in one part of a studio while the show is going on in another part. At present, we find it desirable to construct television sets in portable and lightweight sections without sacrificing sturdiness.

BACKGROUND PROJECTION

The problems of background projection in television differ somewhat from those encountered in motion pictures. More light is necessary because of the proportionately greater incident light used on the sets proper (Figure 4).

Considering the center of a rear-screen projection as zero angle, we must make it possible to make television shots within angles of at least 20 degrees on either side of zero without appreciable loss of picture brightness. This requirement calls for the use of a special screen having a broader viewing angle than those used in making motion picture process shots. Also, in motion pictures, the size of the picture on the screen can be varied to the proper relation to the foreground for long shots or close-ups. For television, the background picture size can not be changed once the program starts. Our background subject matter must also be sharp in detail and high in contrast for good results.

At present, only glass slides are used. A self-circulating water-cell is used to absorb some of the radiant heat from the high-intensity arc. Also both sides of the slide are air-cooled. These precautions permit the use of slides for approximately 30-minute periods without damage.

MAKE-UP

This may be a suitable time to correct some erroneous impressions concerning the type of make-up used in television. It has never been necessary to use gruesome make-up for the modern all-electronic-RCA television system. At present, No. 26 panchromatic base, similar to that used for panchromatic film, and dark red lipstick is being used satisfactorily. From the very beginning, we have made tests to determine the proper color and shades of make-up, keeping in mind that a color closely approximating the pigmentation of the human skin is most desirable from the actor's psychological standpoint.

THE CONTROL ROOM

Now, a few words about the operations in the studio control room during a televised production (Figure 5). All camera operators in the studio wear head-phones through which they receive instructions from the control room. Directions are relayed over this circuit by the video engineer or the production director. Here the televised images are observed on special Kinescope monitors and necessary electrical adjustments are made. Alongside each of these monitoring

Fig. 5—The television control room. Note the two
Kinescope monitors in the upper left corner.

Kinescopes is a cathode-ray oscilloscope which shows the electrical equivalent of the actual picture. Two monitors are provided in order that one may be reserved for the picture that is actually on the air, while the other shows the succeeding shot as picked up by a second or third camera. This enables the video engineer to make any necessary electrical adjustments before a picture goes on the air.

Seated immediately to the left of the video engineer is the production director whose responsibility corresponds to that of the director of a motion picture. He selects the shots and gives necessary cues to the video engineer for switching any of the cameras into the outgoing channel. The production director has, of course, previously rehearsed the performance and set camera routines in conjunction with the camera operators and the engineering staff. The camera operator has no control to switch his camera on the air. All camera switches, which are instantaneous, are made by electrical relays controlled by buttons in the control room. At present, the video engineer's counterpart in motion picture work is the editor and the film processing laboratory.

To the left of the production director sits the audio control engineer whose responsibility is entirely separate from that of the video engineer. He also is in a position to view the monitor, and may communicate by telephone with the engineer on the microphone boom. The audio engineer is responsible for sound effects, some of which are dubbed in from records. His job is somewhat similar to that of the head sound engineer on a motion picture production. Thus, we have the control room staff—three men who have final responsibility for the success of the completed show.

An assistant production man is also required on the studio floor. Wearing headphones on a long extension cord, he is able to move to any part of the studio while still maintaining contact with the production director in the control room during a performance on the air. Actors require starting cues, titles require proper timing, and properties and even an occasional piece of scenery must be moved. The assistant director supervises these operations and sees that the instructions of the production director are properly carried out.

Members of the studio personnel also to be mentioned include lighting technicians, the property man, and scene shifters, whose responsibilities parallel those of their motion picture counterparts. Specially trained men are also needed for operating title machines. In the future all titling will undoubtedly be done in a separate studio inasmuch as operating space in a television studio is at a premium. Today, however, title machines do operate in the studio and require the utmost care in handling. Types of titles used include dissolves and wipes similar to those used in moving pictures.

SOUND REPRODUCTION

As in motion picture work, a microphone boom is used in television production, and is operated in a similar way. Perspective in motion picture sound is accomplished by keeping the microphone, during a long shot, just out of the picture and moving it down closer to the action as the camera moves in for a close-up, thus simulating a natural change in perspective. In television this is not always possible because there are always three cameras to consider. This same condition pre-

-vailed in the early days of motion pictures when it was thought desirable to take a complete scene, shooting both long-shot and close-up cameras, at one time. In the television studio at least one camera is always set for a long shot while the others are in position for closer shots. If the microphone is placed in such a position as to afford a "natural" perspective for close-ups, the succeeding switch to a long shot would reveal the microphone in the shot. You in motion pictures can order a retake; in television broadcasting we can not rectify the mistake. It is quite obvious, therefore, that the man on the boom can not lower his microphone to the "natural" position for each camera shot. We therefore place the microphone in a position just out of range of the long shot. In order to accomplish some sense of perspective between long and close-up shots, a variable equalizer that drops the high and low ends of the spectrum is automatically cut into the audio circuits when the long-shot camera is on the air. In this operation, sufficient change in quality and level is introduced to aid the illusion of long-shot sound perspective. Of course, when a close-up camera is switched in, the audio returns to the close-up perspective quality once more. This may be called remote control sound perspective.

Special sound effects, music, etc., from the studio picked up from recordings are mixed in the control room. In motion pictures, some of the effects and most of the music are dubbed in after the actual shooting of the scene.

The general acoustical problems in a television studio are similar to those in a motion picture sound-stage. Walls and ceiling should be designed for maximum absorption to permit faithful exterior speech pick-up. A stage or studio must be designed to enable presentation of an exterior or an interior scene. With the studio designed for maximum absorption, illusions of exterior sound characteristics can be created. For interiors, the hard surfaces of the sets and props offer sufficiently reflective surfaces to create the indoor effect.

TYPICAL PRODUCTION ROUTINE

After the foregoing discussion of the equipment and personnel, it may be interesting to follow an actual production from the beginning of rehearsal to its final presentation. For this example, assume that we are to produce a playlet (Figure 6). When the scenery has been erected, the first rehearsals begin without the use of cameras or lights. Besides familiarizing the actors with their lines, the rehearsals afford the production director and the head camera operator an opportunity to map out the action of the play. All action, including camera shots.

cues, and timing, is noted on a master script which thereafter becomes the "bible" of the production. Timing is very important because of the necessity of having a particular act time in with the other acts or film subject.

After several hours of rehearsing, the first equipment rehearsal is called. Cameras are checked electrically and mechanically. Focus controls and framing devices are lined up so that correct focus on the ground-glass is also correct focus on the mosaic plate. This completed, the cameras are ready for rehearsal. With the scene properly lighted, the camera operators begin working out movements to pick up the

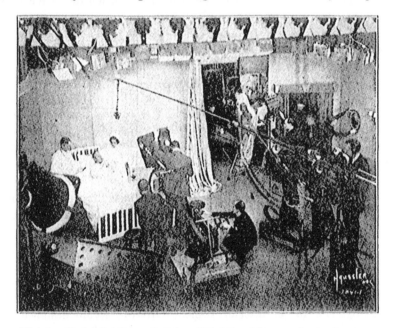

Fig. 6—*(Left)* Scene on the air. *(Right)* setting up for next scene.

desired shots in the proper sequence. The production director instructs the staff and personnel from the control room, speaking over a public-address system. Each shot is worked out and its camera location marked on the floor. At times, the actors may unconsciously depart slightly from the rehearsed routine during an actual show; the camera operator must be prepared and alert to make the best of the situation regardless of all previous floor markings. Continuity is so planned that while one camera is taking the action, another camera is moving to a new location and composing a new shot to be switched on at the proper time. This frees the first camera, which can now move to a third location, and so on. Sometimes during a twenty-minute performance each camera may take twenty different shots. Of course, besides different floor locations, the height and angle of the cameras must be varied to comply with good composition. During rehearsals, timing must frequently be revised to allow for the actual camera movements.

Finally, a dress rehearsal is scheduled. The complete program is televised, including any film subjects or slides that may be needed to complete the program. Frequently the program will begin with a short film leader, followed immediately by a newsreel or a short subject, the film portion of the program coming from the film-televising studio. While the film is running, the live-talent studio is continuously warned as to the time remaining before it must take over the program. Once the studio program goes on the air the production director is no longer able to use the public address system to communicate with the personnel in the studio. Instead, he uses a telephone circuit to his assistant in the studio, and, through the video engineer, communicates by phone with the camera operators.

Another standby warning is usually given when there is one minute to go. Then, as the cue to begin comes, the green light on the title camera is lighted. From this point, continuity must be rigidly preserved. As titles move from one to another, appropriate music is cued in and actors are sent to their opening positions.

With the completion of titles, the image is faded out electrically and cameras are switched to the opening shot. Performers begin their action on a silent cue from the assistant director, who is instructed from the control room. During this first scene, the camera previously picking up titles moves quickly into position to shoot a second view of the action. Again cameras are switched, permitting the first to move to a new position; and so the action proceeds. If the play has several scenes, the concluding shot of the first scene is taken by one camera while others line up on the new scene and wait for the switch. Frequently, there are outdoor scenes. These are filmed during the first stages of rehearsal for transmission from the film studio at the proper time during the performance. The switch to film is handled exactly as another camera switch, except that the switch is to the film studio instead of to one of the studio cameras. The projectionist must be warned in advance to have his projector up to speed and "on the air" at the proper instant to preserve the production continuity. This requires very critical timing, as you can well appreciate. When the film is completed the studio cameras again take over the next interior scene.

Upon completion of the studio portion of the program, one camera lines up on the final studio title, which usually returns the program to the film studio for a concluding film subject.

Since the first program on July 7, 1936, many television programs have been produced. Each has been a serious attempt at something new. Although much has been accomplished, there remain a vast number of unknowns to be answered before it can be said that television's potentialities have been even partially realized. Today, as this paper has indicated, television bears many points of similarity to motion pictures. As a matter of fact, it is likely that television would be somewhat handicapped if it were unable to borrow heavily from a motion picture production technic that has been built up by capable minds and at great expense over a period of many years. Infant television is indeed fortunate to have such a wealth of information at its disposal. Possibly continued experimentation will lead us toward a new technic distinctive of television. During its early years, however, television must borrow from all in creating for itself a book of rules. The first chapter of that book is scarcely written.

TELEVISION

A STRUGGLE FOR POWER

by

FRANK C. WALDROP

and

JOSEPH BORKIN

with an introduction by

GEORGE HENRY PAYNE

Member of the Federal Communications Commission

New York · 1938

WILLIAM MORROW AND COMPANY

1. Prelude to Struggle

SOME PEOPLE LIKE TO SAY THAT TELEVISION IS THE CHILD OF art and science—and so it is, in a way. But, however blissfully art and science may be wedded, the child is no ordinary child, and government and money are deeply disturbed about its future.

To be exact, television represents a synthesis of scientific achievements by means of which electrical analyses of sounds and of the appearance of objects are blended and transmitted in a split second throughout wide areas. Television is just a trick, really; the trick of using electrons in order to look at something not visible to the naked eye. But through the perfecting of this trick the means of access to public credulity, and to the power which that access gives, lie open to some man's grasp—and not enough people know it.

Consider for a moment the report which a group of distinguished Americans, in all seriousness, lately gave to their government upon this matter:

When to the spoken word is added the living image, the effect is to magnify the potential dangers of a machine which can subtly instill ideas, strong beliefs, profound disgust and affections.

There is danger from propaganda entering the schools,

and perhaps much greater danger from the propaganda entering the home. How great is the power in the control of mass communication, especially when helped by modern inventions, has been made clear recently in countries that have had social revolutions, and which have promptly, in a very short period, brought extraordinary changes in the expressed beliefs and actions of vast populations.

These have been led to accept whole ideologies contrary to their former beliefs, and to accept as the new gospel what many outsiders would think ridiculous. The most powerful means of communication, especially for rapid action in case of revolution, are the electric forms like radio and television, which spread most skillfully presented ideas to every corner of the land with the speed of light and a minimum of propaganda labor. Compared with these, the impromptu soap-box orator with his audience of a dozen, or a local preacher with his 200, are at a grave disadvantage. Certainly no advertiser would expect to sell as many goods by amateurish appeal reaching 10 dozen as by a captivating one reaching 10 million.

Television will have the power of mobilizing the best of writers and scene designers, the most winning of actors, the most attractive of actresses.[1]

These soothsayers gave their views to the President of the United States on June 18, 1937, and signed their letter of transmittal, respectfully:

Harold L. Ickes, Harry H. Woodring, Henry A. Wallace, Daniel C. Roper, Frances Perkins, Harry L. Hopkins, Frederic A. Delano, Charles E. Merriam, Henry S. Dennison, and Beardsley Ruml. Even the most casual student of current affairs will recognize among them some astute analysts of ways and means to mold the public mind. As to the scientific import of their study, they told the President they drew upon the National Academy of Sciences, the Social Science Research Council, and the American Council on Education for expert testimony.*

Basic questions of national policy arise. We must know who shall and who ought to control television; what ideas and whose it shall convey. What will be the effect upon human institutions? Who should be rewarded or punished for having brought it upon us?

Perhaps there are no clear-cut answers to these questions. Perhaps there should not be. But at least there are some elements of fact to be had, some definitions of issues, which the citizen can employ as he wishes. The authors of this book propose no answer. But we can hope, and we do hope, that television may not be allowed to fall unknowingly into the hands of some plotmaker, some group in power or seeking power to destroy democracy in this time when man, fretted by his inventions, cannot bear either to throw them away or put them to use according to a conscious plan.

This is an important matter for the common man as well as the special pleader. Communication holds together the very fabric of society, and as social groups grow in power and complexity, their systems of exchanging information become infinitely more important and widespread. Perhaps the most recent major demonstration of the importance of communications to national interest was at Versailles, when the masters of destiny haggled for three things above all others: oil, international communications, and its twin, international transportation. The nations of today are ranked

---

* The National Resources Committee, of which these signatories are members, has made important studies not only of technological trends, but of water power, land uses, and other basic instruments for development of better living conditions.

in power according to their standing in those three cate-
gories.

Television happens to be the newest, and at the same time
the most effective, means of communicating information,
misinformation, and entertainment. To attempt a summary
report of its technical status is dangerous, for scientific standards and conceptions are changing rapidly. Hence we tell
here only what was disclosed or reliably reported as of January, 1938.

The approximate standard performance offered sharp,
clear pictures upon a glass screen seven inches high by
twelve inches wide. Experimenters had succeeded in projecting enlarged reproductions upon screens as great as three
by four feet in area for home use; and some demonstrations
had been given on screens nine by twelve feet, and were improving steadily.[2]

Reproductions were still in shades of black and white,
insofar as routine broadcasts were concerned, but color television was already a practical accomplishment in public display in England and the engineers were busy with even
stranger things. Men were actually searching for the mathe-
matical outlines of ways and means to transmit sensations
of smell and feeling by electricity.[3] Quite soberly, the question was put in the report to the President whether we may
not some day sit in our libraries and have presented to us
the electrical reproduction of distant scenes, in natural colors and sounds, and with every aspect of smell and feeling
except actual substance.

Small wonder, then, that his cabinet officers and their
technical advisers suggest to the President of the United

States that the people of the country should begin to think of what they would like to do with television.

It is far more than a propaganda device for the plausible orator. True, it is invaluable to him. Imagine the candidate for public office standing full length in your living room, pointing to charts, beaming his smile, and exuding the fragrance of roses or stale cigars. Imagine a tottering rule of the elders being restored by some father of his country, long dead but able through electrical reproduction of his living manner to adjure his countrymen once again to avoid evil doctrines and to stray not on strange ways. This is no mad notion. Your phonograph record is a prison for sound. The motion picture film is an imprisonment of sight and sound in one. Grant the engineers their capture and transmission of smell and feeling by electricity, and who will say they can never imprison those sensations, too, for enduring record and reproduction?

But we need not wait until the problems of communicating smell and feeling have been solved to feel the force of this new instrument upon our lives. Sound radio has already indicated the way in which changing technical methods of communicating information affect existing human institutions. Radio, it is by now generally recognized, is a rival of the daily press both for the privilege of distributing news and for revenue from advertisers. In 1937, as the radio industry continued the rapid expansion begun about 1923, newspaper industrialism continued a decline started at the same time.

The trend in radio is indicated by the 1937 financial history of the National Broadcasting Company, the largest sin-

gle program service organization. In that year NBC showed
a revenue gain of eighteen per cent over 1936, and with two
other program companies hired one thousand additional
musicians. Thirty-six new broadcasting stations came into
operation.[4] In contrast, while the newspaper industry, with a
daily publication of 41,400,000 copies, took in $595,000,000
of revenue, it showed a bare two per cent gain over 1936.
There were 2084 daily publications in English in 1937, a
decline of twenty-three from the previous year; 10,629
weekly journals, a decline of 176; and 359 semi-weeklies,
representing a loss of eighteen.[5]

But journals and publications have not yet felt the worst
of competition from radio. One of the most important by-
products of television is the "facsimile recorder," an instrument which will print messages of record in response to
electronic impulses. It can be operated by business establishments to replace telephone and telegraph leased wires
between branch establishments, and also to print news in
the home. The television facsimile machine responds to a
radio signal in the same basic manner that the sound radio
receiving set now does. Indeed, it is designed to be attached
thereto. The electrical impulse causes a stylus to sweep
across plain white paper and bring out not only script or
printed letters but also reproductions of photographs, both
in black and white and in color combinations. Facsimile
machines, operated in conjunction with central broadcasting stations, are today literally capable of producing the
newspaper in the home, eliminating two of the greatest expenses now attached to the publications industry, printing
and delivery. The effect such a radical alteration in methods
must have upon investments in presses, trucks, and build-

ings is obvious. The effect upon employment is equally apparent. Knowledge of these facts may make for understanding of why newspaper publishers are so eager for radio stations, today, and already hold approximately one fourth c all the licenses of operation granted by the Federal Government. They are simply trying to shift their fortunes with the tide of technology.

As in publishing, so in many other trades, industries, and affairs of men. Television already has advanced to such ι state of perfection that it can be, and in some countries is used to amuse. It can present a play in your home just a the drama proceeds in a studio or for that matter in an ordinary theatre. It can report events as they happen. It can fil in the time between by reproduction of motion pictures already caught on film. Mariners have found ways to use certain modifications of television to overcome fog. Soldiers and sailors use it to spot gunfire, and scenes are transmitted from ship to shore, from plane to ground, with clarity and regularity.

Obviously, a device of such powers is not going to be allowed to fall into the hands of one group or another uncontested. Television is an instrument not only of great potential power and uses, but of profit.

Some of the greatest corporate organizations in the modern world are preparing, indeed even now are fighting, to control its development. The American Telephone and Telegraph Company, the Radio Corporation of America, Westinghouse Electric and Manufacturing Company, General Electric Company, Columbia Broadcasting System— these are aristocrats in our financial oligarchy. And they are well aware that if any one of them is allowed to control the growth of television, extinction is threatened for the others.

Inventors are searching passionately for solutions to last details in support of patent claims. Zworykin, Baird, Farnsworth, Finch, Lubcke, Round, Alexanderson, Armstrong, De Forest, LaMert—men whose names for the most part mean nothing to the general public—are the makers of the future for that public.

Not all who are deeply involved in these developments realize what is happening to them. The great motion picture industry and its dependencies, such as the thousands of picture exhibitors, are relatively passive in the face of change. Western Union, Postal Telegraph, and Mackay Radio, in contrast with the great Bell telephone system, seem unable to organize themselves for adequate defense.

But whether they like it or not, television presses change upon them, every one.

2. In the Arena

IT IS ONLY NATURAL THAT THE OPERATORS OF THE PRESENT
sound radio system have assumed that television will gravi-
tate into their hands. After all, they argue, their money has
financed the present development. In some degree they are
right. Furthermore, they are the financial underwriters of
most of the research which eventually must result in obsolescence of the very plant and structure which earns profits
by means of which to endow research. This contention is
less accurate. Corporation engineers appear to have contributed very little to the fundamental development of television. But of one thing there can be no doubt. Unless business men are permitted to swing the basis of financial
development along with the change in technical means of
operation, chaos must certainly ensue.

Radio is no longer an infant industry. Electrical communications operate within the framework of complex and
important corporate organizations. Great fortunes depend
upon right judgment and delicate maneuvering, *and maneuvering now*, for the status of television today is such
that unless the operators of radio, the movies, and several
other industries show considerable speed and intelligence,
they may find their corporate horses shot out from under
them. Television has a popular appeal.

Its progress in England is most commonly referred to, generally because in that nation the government has acted to force operations in which the public can take a part. Widely publicized demonstrations of the reporting of events as they occur, such as the showing of the final tennis matches for the Davis Cup, and the Armistice Day exercises of 1937, caused sensations in the United States where television is not operated so openly for the public. The British programs are sent out from Alexandra Palace, London, and are generally made up of motion picture films which have been exhibited in theaters at least three months prior to the time of televising; of vaudeville skits and dramatic presentations; of "radio visits" to scenes of historic interest and beauty; and of spot news occurrences, such as the Coronation and other State functions. It is highly significant that public interest in British television became intense only after the showing of actual news incidents was possible.

At the beginning of 1937 less than one thousand receivers were in operation, but in December of that year the BBC reported nine thousand licensed receiving sets in daily operation within service range of Alexandra Palace.[1] During the annual radio show, "Radiolympia," sets went on sale at between $178 and $200. The BBC system has been developed on the basis of a government commission's recommendations made in 1935. At that time all inventors and engineers were ordered to place their patents in a common pool from which each could draw the patents of all the others on a royalty basis to build sets and equipment according to his notion of what would be the best.[2] To insure a minimum of good operation, the BBC fixed standards of performance for both broadcasting and reception, with the cost of operation met out of the revenues from taxation of sound radio sets.

BBC has set a standard of performance in accord with the recommendations of 1935, as follows: Programs * are broadcast with a peak power of 17 kilowatts on a 45 megacycle channel for the visual program, and 3 kilowatts on a 41.5 megacycle channel for accompanying sound. The official service range is encompassed within a radius of thirty miles from Alexandra Palace but good reception is reported as far away as Ipswich, seventy miles from the tower. Coaxial cables are being laid from London to Birmingham and other cities to provide provincial service. The pictures are shown at the rate of fifty frames a second, of four hundred and five lines each—very sharp and clear definition.[3]

A public demonstration at the Dominion Theatre in London last January drew an audience of three thousand. Pictures were projected on a screen six by eight feet.[4]

Television received little public attention in France until last year when, suddenly, the Ministry of Posts, Telegraph and Telephone announced that it had ordered the world's most powerful (30 kilowatts) transmitter to be erected in the Eiffel Tower, eleven hundred feet above the earth.[5] British Broadcasting Corporation officials were considerably disturbed, for the French government said it would permit commercial programs, and these, coming from such a powerful transmitter, might easily interfere with the BBC network.

* The reader who is not familiar with technical aspects of electrical communications need not feel concerned. In general, it might be said that when two systems are being compared, the one which is described in terms of larger numbers is the superior. A general analysis of television operation is given within the next few chapters.

The New York Times gave a clue to the conflict within, in a dispatch stating that the French announcement had "stirred speculation in American radio circles whether or not this move augured an international television race comparable to the one now being run in super-power broadcasting." [6] This conflict was adumbrated as early as 1933 when a British program was broadcast to a theater crowd at Copenhagen.[7] Undoubtedly, as technical proficiency advances some compromises will have to be made on television frequencies for international broadcasts along the lines now existing in sound radio. The French station is four times as powerful as any now licensed in the United States. No data are available in regard to picture definition.

In Germany television is being operated very efficiently by the Post Office Department. Exactly how many transmitters are in use is not known, but at least one is in Berlin, broadcasting on a radius of sixty kilometers of service range. Another in the Harz mountains has a radius of one hundred and twenty kilometers.[8] Five companies are manufacturing sets with large screen, cathode ray projectors, with picture definition of one hundred and eighty lines, twenty-five frames per second.[9] Very successful work was done during the Olympic Games in catching action scenes. One of the most interesting German developments is that of public television-telephone service, by means of which a person may see the one to whom he is talking by wire. For some time such a service has been maintained between Berlin and Leipzig, and last year the government authorized extension to several other cities.[10]

Russia is reported to have erected television transmitters of low caliber definition in Moscow, Kiev, and Leningrad, but apparently intends to go in heavily for further development, probably for military purposes. An order was placed with the Radio Corporation of America in 1937 for a 7.5 kilowatt station costing one million dollars.[11] An unofficial report stated that RCA was also retained to make receiving apparatus and that contracts were made for the use of RCA patents.[12]

In Italy, SAFAR, the authorized manufacturer of television equipment, reports technical efficiency on a standard performance of twenty frames of three hundred and seventy-five lines each per second. A chain of stations connected by coaxial cables and operating on service ranges of twenty-five miles radius each is reportedly being considered by the government.[13]

As early as 1932, the Japanese Radio Broadcasting Association claimed to be able to exhibit pictures on screens eight by twelve feet, but of undisclosed definition. Stations were being maintained by the government at Waseda and Hammatsu Universities, but so far as can be learned they were of narrow range and low definition.[14] Since then, considerable progress appears to have been made. It is expected that by 1940, when the Olympic Games are scheduled to come to Tokyo, a public broadcasting system will be in operation covering a twelve mile service radius.[15]

In Poland, Czechoslovakia, Holland, and Sweden television experimentation was progressing last year. Each of these countries has a transmitter offering experimental programs to the public, but no original research or basic patentable discoveries were reported.[16]

The important characteristic of television abroad is that in every country it is being conducted by government departments or in close co-operation with the government. In most cases, direct governmental subsidy underwrites the laboratory work and the public operations as well.

In the United States television is unquestionably more advanced, with respect to station operations and technical efficiency, than anywhere else in the world. There are eighteen licensed stations; and because so little is known about them by the general public, we have listed them in detail.*

What are the powers of television? It permits the leaving of messages. It is powerful in scattering persuasive arguments among masses of people. And the same electronic means that produce television permit multiple telephone conversations by radio or wire. And this is not all. Television informs, entertains, spies for gunners, guides mariners, prints newspapers. And not only the publishing industry may anticipate corporate corrosion as a result of its workings. Consider the case of the amusement trades. The clown never got rich from one performance. He collected his pennies in weary travel from village to village. The stars of Broadway grew proud as the lines grew long in front of box offices, for they knew that long lines meant long runs, long lives for plays and work for actors.

The movies changed all that. The clown's humble repertoire made just one short reel of laughter and was gone forever. He traveled no more, but let the can of film do the trouping. Broadway's stars and what has happened to them start no tears today. Everybody knows how Hollywood has ruined one Broadway and set up a thousand others across the nation until now we have a Broadway wherever there is a marquee and a billposter proclaiming next week's drama.

* See Appendix A.

But not everybody knows what is in the making for Hollywood. Suppose the clown's short and simple annal of amusement is presented to the whole nation in one brief moment. Its travels are over, once the nation's television sets have flashed the antic to all of America's homes in fifteen minutes flat. And travel is ended, too, for the great feature film. Why struggle downtown through traffic, then stand in line, and pay money to see *Mutiny on the Bounty*, when it can be enjoyed at home just as well?

The unhappy newspaper publisher, too, finds that by installing facsimile printers in people's homes to escape the expense of operating printing presses and delivery systems he only adds other burdens elsewhere. He cannot junk his machinery, turn the workmen out into the street, and go singing on his way. People demand support, whether or not they labor. In the United States, at least, a press free of censorship is guaranteed by the Constitution. But can the facsimile be called an instrument of the free press? That is an issue not yet settled for the publisher and he cannot face it with any certainty of success, for the Government holds firm control over facsimile's common carriers, as well as over those of all other electronic devices of communication. Moreover, the Government is aided in keeping its grip by the confused state of all thought concerning relationships between technology and human institutions. Not even in law can any basic definitions of private rights involving electric communications be given with certainty.

"Notions of sovereignty, states' rights, property, laissez faire, developed by land and commercial economics, are belied by the scientific facts of this novel method of communication," says the *Encyclopedia of Social Sciences*. "The peculiar characteristics of radio have evoked a distinct radio law, but the legal controls have been shaped largely by the state of the art, and require continual revision if they are to keep pace with its progress."

And we shall see, presently, that law has never been able to keep pace with art. Neither has government control of propaganda. One characteristic of the radio technology remains: if a program is broadcast, there is no way of being sure the wrong people will not hear it. This matter must be solved before television is as common as sound radio, or the absolutists are lost.

The untrammeled electron is at once a pleasure and a pain to the politician in power. By means of it he can address a whole people, but by that same means a whole people can be reached by his competitor if that competitor can gain access to the transmitter. In countries where absolutism is supposed to be the order of the day, the dictator commands his victims to attend the radio as faithfully as the Moslem heeds the muezzin's call to prayer. But he is tortured by the knowledge that some scoundrel from beyond the border, or even within it, is likely to commit piracy upon the sanctified domain and reach startled ears with unsanctified information.

To maintain their grip upon mass sentiment, indeed to forbid the exercise of intelligence, the iron men have been driven to some extraordinary measures. Japanese police, for instance, are charged with the responsibility of eradicating "dangerous thoughts." In Germany revolutionary suggestions from outside have become so common that the government is reported as planning a most extraordinary attempt to "save" the people by taking radio, as we know it, out of use. Early in August, 1937, there was an exposition in Berlin of high frequency radio developments, and that was very important. It revealed the purpose of changing the entire system of radio in Germany from wireless transmission to transmission by cables. The value of such an arrangement for war purposes is obvious; it would be safer from interruptions and against destruction. Besides, it is easier to control the programs of listeners, and to prevent the reception of "subversive" programs from other countries [17]—or even one's own. In March, 1937, programs being broadcast in Germany could be heard plainly in New York and Pittsburgh. Each period opened with a singing of the "Internationale," amounted to a harangue for development of a united Socialist-Communist front, and closed with the hymn of revolution. Hitler's agents combed the Fatherland, but if they ever found the daring broadcasters the world has not been told. The station happened to be mounted on wheels.[18]

Hitler's resolution to abandon radio broadcasting in favor of wires for both sound programs and television touches upon a basic problem of the whole industry, that of monopoly. But before we consider it, let's worry some more with the administrators of government. The cross-fire propaganda between warring dictators in Europe is a common topic of political conversation. Only lately have the gossips become aware that the cross-fire is no longer confined to one continent. Approximately forty programs are being broadcast from Europe daily in foreign languages and in English translations, intended exclusively for listeners in the Western Hemisphere. The Congressional Record offers as exhibits the mailing lists of American branches of the great salvation systems. To the names on these lists are sent cards every week advising the comrades at what hour and upon what radio frequency to heed the words of wisdom from afar. Toward the close of the year, broadcasts from foreign countries were arriving steadily not only from Europe but from Asia, Australia, and, of course, South America.

Nor was the Government of the United States allowing these extra-national campaigns to proceed unchallenged. For domestic consumption, the Department of the Interior's Bureau of Education organized a series of programs characterized by an opening hymn entitled, "Let Freedom Ring." Officials of the administration were found to rush before microphones at the slightest opportunity to explain every minor matter of policy, and the President of the United States was considered a professional master of the art of talking to a nation from the fireside.

For international consumption several powerful, privately financed stations were in operation. Their programs, however, were offered only after approval by the United States Government and were directed chiefly toward Latin America, in conformity with the "good neighbor" policy of the administration. Terms of highest praise for democracy, liberty, and other techniques of libertarian government considered disreputable elsewhere in the world were common characteristics of these broadcasts. The policy of allowing private interests to distribute these programs was distinctly declining in State Department favor by the fall of 1937. One bill pending in Congress proposed authority for the erection of a tremendously powerful government owned station directly dedicated to broadcasting of pro-American propaganda to the world.[19] Government departments clamored for equipment and broadcasting powers for propaganda, message exchange service, and for entertainment, pure and simple.

The question now is whether trends set up in sound radio will not prevail with television. The great technical issue of today is clearly indicated in the reference to plans for putting all German communication back into wires. Hitler's objective, naturally, would be to exclude interference of the sort that comes in pure radio operations. Every feature of television can be offered by wire distribution, except, of course, when it is used in connection with vehicles in motion. Shall communication in America be by wire or wireless? It is not entirely a simple matter of engineering technique. If television is confined to wired services, it is likely to continue expensive and therefore difficult to put into common service. If it is broadcast after the sound radio principle, then great areas of the nation may never receive it. Furthermore, a tremendous enterprise, that of radio operations in general, must be revised.

The Federal Government is being burdened with the multiple task of setting standards of performance, deciding between contestants for the right to perform, enacting legislation which will preserve all equities, and repelling political boarders, as it were, who seek to use sound radio and television to contaminate our institutions. Before we can guess what the government can do about it all, we must understand something of television's scientific structure.

8. Trouble in Heaven

ONE PORTION OF THE SPECTRUM IS VERY NEW. IT IS KNOWN as the ultra-high frequency segment, ranging from 30 megacycles ad infinitum, and it is the field upon which radio's titans are gathering for a tremendous struggle. In October, 1937, the Federal Communications Commission announced it would consider applications for licenses in the zone between 20 and 300 megacycles, and indicated its feeling in the matter with a little homily:

The allocation of the ultra-high frequencies vitally affects several important broadcast services, namely: television, facsimile, relay, high-frequency and experimental broadcast services.

The action taken by the Commission today with respect to television is merely one step of many which are required before television can become a reliable service to the public. Some of these many steps must be taken by the industry in the development of proper standards which in turn the Commission must approve before television can technically be of the greatest use to the public on any scale.

Also the Commission, at the proper time in the future, must determine the policies which will govern the operation of television service in this country, particularly with reference to those matters which relate to the avoidance of monopolies. And the Commission must also in the future prescribe such rules and policies as will insure the utilization of television stations in a manner conforming to the public interest, convenience and necessity, particularly that phase which will provide television transmission facilities as a medium of public self-expression by all creeds, classes, and social-economic schools of thought.

The investigations and determinations of the Commission justify the statement that there does not appear to be any immediate outlook for the recognition of television service on a commercial basis. The Commission believes that the general public is entitled to this information for its own protection. The Commission will inform the public from time to time with respect to further developments in television.[1]

After such a statement the commission can never plead ignorance of the issues. But what shall we make of its behavior? We know there is no question about the technical efficiency of television. Promoters and engineers are agreed that America leads in the technical aspects of all forms of communication by electricity. The real problem unquestionably is one of resolving conflicts between applicants for permission to perform.

Here is an example of the sort of mood in which these people approach the governing body:

I wouldn't have the intestinal fortitude—plain guts if you would rather have—even though representing the important police service I do, to stand before you in an attempt to confiscate the important band between 30 and 42 megacycles to the exclusion of commercial and other interests who have just need for such channels and for promoting the public good and welfare.

And if any service, governmental or otherwise, think they are going to get away with this without hearing from the police service which protects the lives and property of civilians in times of peace as well as in times of war, they are sadly mistaken. . . .

While a thug stands with drawn gun and cocked hammer we would betray a sacred public trust if we didn't seek our just share of frequencies and we are not going to be hoggish about it, either. . . .

We have no paid lobby but we do not intend to draw our punches for the benefit of the thug and to the detriment of the public at large.[2]

This is just Captain Donald S. Leonard serving notice, on behalf of the International Association of Police Chiefs, that these gentlemen, operating on the lower bands of the spectrum, want a place up where television may sprawl. He is indicating, rather melodramatically, the belief that police radio serves the public interest, necessity, and convenience sufficiently to warrant its continuance and expansion.

And William S. Paley, president of the Columbia Broadcasting System, holds that if private capital is going to continue doing the sort of broadcasting job it has started out to do in this country, its past investments must not be ignored.

I say this because there must be constant encouragement to capital flow if the people of America are to have the benefit of every technical discovery, every creative advance. For this reason, sudden, revolutionary twists and turns in our planning for the future must be avoided. Capital can adjust itself to orderly progress. It always does. But it retreats in the face of chaos.

We are on the threshold of a period of transition for the next couple of years. We should do everything in this period to advance experimentation. But we should do nothing to weaken the structure of aural broadcasting in the present band [of the spectrum] until experimentation in other bands has yielded to us new certainties.

For instance, allocations in the present broadcast band are such that even a few minor changes might upset the whole plan of the structure. The present layout is like a chess game. A single move can have almost infinite ramifications.

Probably the most important economic problem we must face—certainly the one uppermost in everybody's mind—lies in television.[3]

Not long after that declaration of his views, Mr. Paley made an extremely forehanded move in the interests of his company which is, on the whole, just a program service, with the duration of its life dependent upon the licenses of Columbia's outlet stations. On June 7, 1937, he filed with Securities and Exchange Commission at Washington an application for permission to sell shares to the general public in its going concern. The acceptance value may be judged from this: to initiate its chain of station outlets, the Columbia Broadcasting System expended in cash $1,600,000. The stock issued in June against this enterprise was sold to the investing public at market prices indicating a potential gross return of fifty-five million dollars upon the whole issue.[4] Mr. Paley has a reasonable right to assume that those investors will join him in an alert interest in any readjustments of the radio spectrum which might endanger their investment. If any of them had any fears concerning the six months' license provision, it was not recorded.

How does television imperil these vested interests of which Mr. Paley speaks so tenderly and Captain Leonard so vehemently? Here's an example: a "shadow," or unexplained interference, upset commercial and all other broadcasters along the Pacific Coast during 1935–36. For quite a while the scientists argued seriously whether or not the mysterious activity was not the long predicted message from Mars. Finally, it was discovered that the "message" was coming from diathermy machines with which doctors treat syphilis, arthritis, and give simple pleasure to hypochondriac movie stars.

V. Ford Graves, chief inspector of the Federal Communications Commission's western division, estimated that of the fifty thousand diathermy devices reported in use then by the American Medical Association, some forty-nine thousand were buzzing away in California. Now the most common (1935) model shoots heat into the human body by high frequency radio current of the 6-20 megacycle variety, but of relatively low volume. Newer types are rising both in power and in frequency, to threaten interference with radio activity in the entire upper area of the spectrum. No medical license was required either to make, own, or operate these increasingly popular instruments, as of 1937. Mr. Graves reported that one private citizen in Los Angeles, not a doctor, operated eighteen of them all day long to the interference of all forms of upper-band communication, even that of the United States Navy on maneuvers off San Diego.[5] One can imagine the howls that would arise if these fascinating titillators of the arthritic were limited by government fiat to use on rigid schedules. And yet, if the doctors are not restrained they may eventually blanket and scramble the radio communications systems in daily use, and send flickering distortions and shadows across the television screens of America to bring about a headlong collision of interests between the sick and well.

But any possible trouble with doctors would be mild compared with the existing conflict outlined by Dr. C. B. Jolliffe, who resigned the post of chief engineer of the Federal Communications Commission to take a similar position with the Radio Corporation of America, one of the chief practitioners before the commission. He declares that the quality of a television picture is rigidly determined by the number of picture elements. The number of picture elements determines the frequency band which must be imposed on the radio frequency carrier. There is no short cut and no compromise. Consequently, we must face the fact that good television requires a wide band of frequencies. Good television can be included in a band width not less than 6 megacycles, but reduction in that band width will reduce the quality of the picture which it is possible to transmit.

When one considers the fact that all of the commercial auditory radio in the United States is jammed into an area on the spectrum between 550 and 1500 kilocycles, one can realize just how radio engineers and investment operators feel about the presence of "the great gobbler," television, in the zone just opened for licensing.

Dr. Jolliffe's doctrine is that reasonably good television can be broadcast over that area of the spectrum between 42 and 86 megacycles, but that each such broadcast must consume, vertically, 6 megacycles of spectrum space. But is that the whole case?

Say that a program is televised on a frequency of 42-48 kilocycles: would it then ripple across the continent to be picked up in San Francisco by really good receivers as easily as in New York? Not in the present state of the art, says Dr.

Jolliffe. The current "horizon" or perimeter of reception for visual broadcasts is about forty miles from the transmitting station. But the effects are not so limited, for interference and incoherent radio activity reach out on a radius of two hundred miles from the antenna.

Given the area between Boston and Washington to be served by television, Dr. Jolliffe works it out thus:

1. The distance between the two cities is, roughly, four hundred miles. Therefore, a station at Boston and one at Washington may emit programs on equal frequency, safe from effect upon each other's audience.

2. To send the same program from Boston to Washington by radio entails the use of "booster" stations, erected every forty miles to catch the program on one frequency and toss it on to the next station on a frequency of different register, which would interfere with no other station within a radius of two hundred miles. One quickly sees that the use of "booster" power involves complete exclusion of competition in the ordinary sense.

For if the booster receives on one frequency and emits on another, to escape interference within a two-hundred-mile range, and yet boosts the image only forty miles, the next station must likewise consume a third segment in order that its rebroadcast escape any interference. Each station carries over one frequency as it were, and picks up one new one. A single program sent from Boston to New York by booster power therefore consumes the whole series of television bands between 42 and 86 megacycles, and demands monopoly if Boston, New York, and way-stations are to enjoy the same program at the same time. Television, if it is to be successful, must approach universality of acceptance among the people. But must acceptance be at the price of its spectrum space given to a single operating concern? [6]

Dr. Jolliffe did not say so, but there are other means to disseminate television programs. A device owned by the American Telephone and Telegraph Company known as the "coaxial cable" is now in operation. By means of it, the television impulses can be propelled not forty, but as many hundred miles as one may wish between broadcasting stations. The broadcasters need only observe the law that stations within two hundred miles of each other broadcast on different frequencies. That is a state of affairs common to aural radio. And that is a state of affairs still involving monopoly. Not monopoly of the spectrum, it is true, but monopoly nonetheless; the sort of moral ascendancy the American Telephone and Telegraph Company now has over sound radio and sound motion pictures. It is necessary to understand the telephone company's interest in the radio spectrum if one is to appreciate matters of discussion further along.

In the zone between 1.6 and 30 megacycles, all of America's domestic telephony is laced by radio to ships at sea, the wire networks of more than sixty foreign countries, and airplanes in flight. As the spectrum exploitation jumps to 30 megacycles and above, the telephone system's interest jumps smartly along with it.

Lloyd Espenschied, radio transmission development director of the Bell Telephone Laboratories, Inc., states the prospects of his organization in this upper region to be of greatest importance. Two way service between ships, planes, and motor cars can be expanded and revised. Doctors can be called while driving in the country, can answer and ex-

change information. Armies in the field and navies on ma,
neuver can function in closer contact with headquarters if
they can shield their machines from interference and their
messages from interruption.

‾In point-to-point service, the number of circuits that can
be developed for simultaneous use can be vastly multiplied.
Mr. Espenschied stated, in fact, that the upper megacycle
radio channels and the Bell system's vitally important new
coaxial cable have similar characteristics.[7] The cable will accommodate more than four hundred telephone conversations simultaneously, or one television program involving
as many lines of definition. Thus we can see that the Bell
system has great interest in the future of the radio spectrum
above 30,000 kilocycles. Coaxial cables are priced at four
thousand dollars a mile.[8] Radio channels cannot be estimated in terms of depreciation, upkeep, repair, nor, so far,
of taxation. If a great network of cables is built, but not required for use in television, what becomes of investment
and income?

Of course, the outlook for substitution is, as Mr. Espenschied is careful to claim, subject to current limitations of
engineering powers. But, as he is equally generous to admit,
there is no basis for assuming that engineering powers are
even temporarily halted in the advancement of wide general uses of the upper megacycle zone.

The important thing to keep in mind is not that engineering is still pioneering in this area, but that engineering
indicates a transmission band of 6 megacycles for the type
of telephony Mr. Espenschied discusses and an equal band
for television. It all works out very nicely from the engineering standpoint. But what about the telephone ratepayer?

Does he have to worry about the clearly implied conflict? Not, of course, if he has no complaints against the cost of service.

But the representatives of other, unsuspected, interests are not so casual. Geophysical prospectors who sound the inside of the earth for gold and oil and copper and iron by high-frequency current want to know the future of the radio spectrum. It means dollars to them. And it concerns the efficiency of government, too. Dr. J. H. Dellinger, of the International Bureau of Standards, demanded of the Communications Commission more than half the available frequencies between 20 and 192 megacycles on behalf of radio-using bureaus of the Federal Government. He got what he sought but not without limitations, for the agencies which keep them do so at the risk of being accused of that worst of crimes, "government competing with business"; and against the will and effort of many a person within as well as without the Federal Administration.[9]

Shall doctors treat syphilis and cancer to the detriment of naval communications? Shall television be set aside in the interest of field maneuvers of a tank corps in the Kansas prairies? It's everybody's problem.

The American Telephone and Telegraph system wants to expand, naturally. So (and it is no secret) does the Radio Corporation of America. Each sees the advantage of high-frequency transmission through the use of booster power stations as described by Dr. Jolliffe and admitted by Mr. Espenschied. And though they are agreed now not to fight, agreements have a way of fading before the necessity of self-preservation. Shall agreement in this case fade in the interest of wires or waves? The decision rests not with the contestants but with the referee, who has set about some-what timidly to test his strength.

He has apportioned seven channels for television in the spectrum band between 44 and 108 megacycles, and twelve more between 156 and 300 megacycles, each channel 6 megacycles wide and providing for picture and synchro-nized sound. For old-fashioned sound radio, seventy-five channels are made available between 14.02 and 43.98 mega-cycles, and twenty-nine special police broadcast channels are provided between 20 and 40 kilocycles. Further provi-sions are made for aviation, geophysics, fixed point-to-point forestry, marine, and all the other familiar subdivisions of interest we met down at the lowest levels.

What stirs the blood of the radio man is the commis-sion's announcement that applications for licenses must be filed with it before October, 1938, for allocation on a defi-nite basis in 1939.

12. The Somnolent Cinema

TELEVISION EATS UP LARGE AREAS OF THE SPECTRUM TO THE starvation of other radio services, but that is not the end of its ravening. It threatens to swallow whole industries. Radio set manufacturers will have to transform their technique of production so that they become television set manufacturers. Radio broadcasters must become television broadcasters.

The radio set manufacturers, and the broadcasters who found the commercial band of the spectrum a vein of virgin gold, have recognized full well the danger that confronts them. In regiment formation they have bombarded the Federal Communications Commission to consider their interests as television approaches. They experiment, make treaties among themselves, and offer plans for protection. They might be called sprinters, crouched for the starting gun in a race that will end in fame and fortune for somebody. But among the contestants we see an unwilling fat boy trying to assume the angular position of the ostrich with head in sand. That, in a word, is the way the motion picture industry is behaving as television comes. The bulk of television programs will probably be in the form of motion picture films. For one thing films are more easily televised than stage performances, and have proved so successful that in the present experimental period sixty per cent of the broadcasts are from films. Apart from mechanical perfection there are other considerations. The film story technique lends itself naturally to television; and so does the scenic perfection that the motion picture industry has developed.

But television has a voracious appetite for material. If it comes to operate on a time schedule equal to that of present commercial radio, the present annual production schedule of films will not maintain service for more than three months. To keep up with such a pace the movies will have to undergo radical changes. Present production schedules, if quadrupled, still would not meet the demand. But even if the supply of entertainment can be kept up, the movies may still be reduced to a minor vestigial program service unless a sound bargaining position is established for them. Having undergone one radical change in ownership and financial structure because of unpreparedness, the movie moguls ought by now to be alert to technical change and its threats, but, alas, they seem not to be.

At present the motion picture industry is in two distinct though not entirely separate branches, each dependent upon the other. One branch is concerned with the production and distribution of pictures (Hollywood), the other with exhibition (America). Hollywood concerns itself with studio operation, photography, sound recording, the selection of artists and plots; in a word, with picture creation. Production could go on in a television era, only speeded up or slowed down to meet demand; and nobody outside Hollywood, except those holding stock in movie companies, would know or care.

The exhibitors simply put the finished products before America today and try to ward off the headache which is surely going to overtake them with the advent of television. It would appear as though, when the new consumers are available at the studios, the producers may be in a measure freed from their dependence upon the exhibitor to whom they have had to cater for so many years; but actually the television broadcaster is merely substituted for the exhibitor.

The movie moguls have always been the victims of a mania for, and a complete failure to attain, independence. Before the advent of sound they used their fresh and copious profits to create exhibition outlets of their own wherever possible. Some of these remain today. One of the first ventures into both sides of the market was made by William Fox, a furrier turned nickelodeon operator who acquired a producing company to guarantee his theaters films for exhibition. Fox is a rare character and one of those who make this story possible, for he not only bound production and exhibition together, but overlaid both with sound and with banknotes. At the advent of sound, Fox intensified the chain movement of theaters by pushing the industry into the new technique so that it had to be assured not only of actual distribution of product, but also of equipment in theaters to reproduce programs in a manner becoming to the super-colossal empire that Hollywood conceived itself to be. On the practical side it was recognized that the movies could not go on half silent and half sound. Events and schemes pressed the moguls finally to choose sound.

The arrival of sound movies smashed the structure of such leading companies as Fox, Universal, Paramount, and Radio-Keith-Orpheum, and made them the vassals of bankers. Famous actors and actresses became as obsolete as wooden plows or handmade shoes. Theater orchestras vanished into picket lines; and the legitimate theater became an appendage. Today those few actors who refuse the western adventure find themselves cast in productions which are conceived, designed, and maintained in the sole hope that some film company will take an option on them. Is it inconceivable that the next step in the theater's metamorphosis is a vestigial movie house in which to test public reaction before the great exhibition to the nation by way of the radio spectrum? Will the motion picture theaters occupy the present situation of the legitimate theater? To determine such questions as these the movie industry maintains an institution known as "The Motion Picture Producers and Distributors of America," headed by Will Hays, who was Postmaster General of the United States during the administration of Warren G. Harding.

In 1936 Mr. Hays hired A. Mortimer Prall to make a study of the relation of television to the motion picture industry. Upon learning that this research student was the son of the late Anning Prall (who was then chairman of the Federal Communications Commission, which also had the problem of television under study at that time),[1] one recognizes the astuteness of the "Czar of Hollywood."

Mr. A. Mortimer Prall, in a highly confidential document entitled "Television Survey and Report," advised the movie people that television opens a new and extremely important field for the industry. He pointed out that three times the amount of film they produced would be necessary for television. In addition, "the motion picture industry is composed of great production corporations. They possess every

94

element necessary to the production of the finest programs of sight and sound on film. Writers, composers, artists, designers, architects, engineers, technicians, construction men, studios, special equipment and the world's best actors and actresses are all part of this industry . . . It is clear that the motion picture industry is the only source of supply for television programs."

Two plans were suggested in this report. One was that the present producers apply to the Federal Communications Commission for permission to buy up one of the existing radio chains such as National Broadcasting Company, the Columbia Broadcasting System, or the Mutual Broadcasting System. The other was that the motion picture industry buy up stations not now in one of the four major networks and form a fifth radio chain. That too necessitates application to the commission for license. In other words, he suggested that the motion picture industry engage in the business of radio with the sanction of the commission of which his father was chairman.

There are several obvious faults in this plan. Sound radio is certainly a step towards television. But it must be recalled that television will play in the upper strata of the spectrum. There is, of course, no guarantee by Mr. A. Mortimer Prall that the commission will give the movie industry frequencies for television when the day for commercial exploitation arrives. It could happen that the movie industry would find itself left with two very large and moribund white elephants —the present motion picture studio and theater system, and the sound radio system as well.

Is the exhibitor to be left to his fate by Mr. Prall? This is an important consideration, both for the producers and for the little men with neighborhood theaters. Because of their large investments in exhibition chains it would be suicidal to their capital structure for the great producing systems to allow their theater investments to crash. But however we may pity them we have to ask what incentives there will be for a customer to drive his car, run or even walk to a movie house when his own living room may become a theater; and we can think of none that seems valid. Maybe there are reasons why the movie palace will last despite television. One argument has been advanced to the effect that the theater will remain as a place of assembly because man is naturally gregarious, but that possibility seems a poor comfort to the magnate whose fortune has to depend on it. Rather, he turns to a report of the Academy of Motion Picture Arts and Sciences which differs with Mr. Prall absolutely. It states that all is well and that the motion picture industry has nothing yet to worry about from television.

"There appears no danger that television will burst unexpected on an unprepared motion picture industry," [2] says the Academy, and since that is comfort from his own, the magnate dreams comfortably of *apfelstrudel* and dividends. Whether this is simply whistling in the dark, or is a private word of assurance based on evidence undisclosed to the public, is anybody's guess; but at the risk of destroying peace of mind in Hollywood, we offer as a clue the following clause for a contract that conditions production by ninety per cent of the sound motion picture industry:

No licenses are herein granted or agreed to be granted for any of the following uses or purposes:

(1) For any uses in or in connection with a telephone, telegraph or radio system or in connection with any apparatus operating by radio-frequency * or carrier currents. . . .[3]

Television can operate only on radio frequencies, or on carrier currents through wire cables. This clause is a part of the contracts between the American Telephone and Telegraph Company and seven of the eight major producers of pictures in Hollywood. Have the movie men been assured by their masters that television will be allowed to develop only as the masters will? Or have they overlooked that clause entirely and simply concluded that movies have their place in the world and can't be shaken out of it? We cannot but succumb to our habit of quoting official documents as a means of showing that there is more than guesswork and intuition behind the warning that the movies may be on their way to extinction or absorption. Bear with us in a flashback of history concerning the sad story of the silent film and the sound machine. It is told briefly in two excerpts from the memoranda of a memorable character whom we shall identify shortly. He, more than any other, drove the nails in the coffin for Gene Fowler's fabulous "Father Goose." Here is memorandum number one:

The motion picture industry in the United States owes us about sixteen million dollars and our expected revenues from the industry for the next ten years is about sixty-five million dollars. This is a large stake and establishes our interest in the welfare of the motion picture industry.

* In the first sound recording contracts between the Bell telephone system and the Vitaphone Corporation, television was specifically mentioned, but in characteristic fashion this was withdrawn as events and legal stipulations came near toward conflict.

The industry is in a serious financial condition and some of the large companies are faced with possible receiverships. The morale of the management in many instances has been greatly lowered. Unwise remedies are being applied and re-organization efforts are being made that in all probability will not be successful. As a result of these conditions our stake is in jeopardy.

We are the second largest financial interest in the motion picture industry. Our stake is next to that of the Chase Bank. . . .

I believe that the protection of our interests in the motion picture industry requires that we should have authoritative conferences with the Chase Bank at the present time. Our interest should be made clear and our influence felt. We can do things the Chase cannot do in the interest of the common good and Chase can do things we cannot do. . . .[4]

Number One was written on November 5, 1932.
Number Two:

I have also had innumerable proposals that ERPI go into this or that phase of the motion picture business. These I have declined without bringing to your attention because I recognize such proposals to be contrary to the Bell system policies and interests, and even though they offered ERPI opportunities for advantage and benefit. It is true today, as it has been for three or four years, that the Telephone Company can control the motion picture industry through ERPI without investing any more money than it now has invested.

I am not recommending that this be done, even though I know that the salvation of the picture industry lies in this direction. The industry is in crying need of the kind of strength and character that could be obtained through the influence of the Telephone Company.[5]

Number Two was written December 7, 1933.

Had "this direction," as described in the correspondence between J. E. Otterson and E. S. Bloom, officials of the American Telephone and Telegraph Company, been followed, all of the motion picture industry would soon have found itself under a single management, with a single studio operating organization and turning out pictures to be sold and exhibited through apparently competing sales systems. And, according to most standards of artistry and theatrical enterprise, disastrous effects upon the movies as entertainment would have been invited thereby.

It is crystal clear that only the judgment of its distant financial masters left the motion picture industry a figment of independence when it tottered under the impact of sound technique. That figment of independence has been nourished carefully since, but never enough to allow the original moguls to re-establish themselves completely.

Let us remember and never forget that of the eight major producing companies, seven are bound up so that they cannot sell or lease their films for television if they want to; and that is why, perhaps, the Academy of Motion Picture Arts and Sciences recommends no fears. They put their faith in the cool judgment of the financiers far away to ward off the new threat. But what of the eighth major producer? And what of that great industrial magic, Competition?

The telephone system moved in on the motion picture industry with a new technology, the sound films, and tied up ninety per cent of production with its contracts. Of the remaining ten per cent, the apparent competitive fringe, virtually all fell into the hands of the Radio Corporation of

America, which proposes itself to be the perennial nemesis of the wired communications services.

And not too unsuccessfully, as witness this further memorandum by an A. T. & T. Company official:

> In the talking motion picture field they [RCA] are competing very actively with us at present, as you know, to develop an affiliation with the large motion picture producers, and competition between us all will doubtless ultimately result in a situation highly favorable to the motion picture interests and opposed to our own.
>
> This is an extensive and highly profitable field and it is quite worth our while to go a long way toward making it practically an exclusive field. I believe that we could justify from a commercial standpoint paying a large price for the liquidation of the Radio Corporation for this purpose alone.[6]

The author of this remarkable view was by no means foolish. Events show that he saw correctly the problems of protecting vested interests in times of technological change. And perhaps it is because the motion picture producers realize that they are really in no position of command just now that they cower like white rabbits as events start their march again. But what about the movies' masters?

19. Past Is Prologue

WHAT IS THE VALUE IN REVIEWING THE PAST?

De Forest has faded from competition with RCA, which bought up his bankrupt plant, and most of his colleagues have gone the same way. The Bell system has settled its differences with RCA in sound motion pictures and neither offers to compete with the other in broadcasting or in domestic telephony, but television is another matter. The amended treaty of 1926 fails to dispose of it in a clearcut manner, and we know that the acrimonious exchanges of Mr. Jewett and Mr. Sarnoff concerning the respective values of the coaxial cable and the radio spectrum indicate that neither intends to allow the other to dominate.

Where stand the remnants of competition? What may be expected of that people's champion, the Federal Communications Commission? Is the struggle for television to be another exhausting battle such as that which we have recounted in sound radio and sound motion pictures? If the radio were not so intimate a force with the American people, or if the American people were more intimate with the forces that have controlled the development of radio so far, it might be unnecessary to have any concern for these matters.

Of three business institutions with most at stake, we find one somnolent even though warned by a Paul Revere who certainly could not be said to have detoured headquarters. Such is the case of the motion picture industry, which shows no apparent interest in the report on television given it by A. Mortimer Prall, whose father was the chairman of the Communications Commission. There is nothing somnolent about the Bell system. It is divesting itself of the sound motion business by selling ERPI and settling numerous anti-trust suits by independents in that industry. Like a champion boxer, it is poised for action. A cable is already laid between Philadelphia and New York, and rules of service for television and telephony have been made, one of the most interesting of which is the Communications Commission requirement that wires be used instead of wireless for relay of programs wherever that is at all possible.[1] And there is nothing somnolent about the Radio Corporation of America. If anything, its conduct is feverish. Unlike the Bell system, it has failed to soothe those whom it has been unable to destroy, and it has failed to destroy some who thirst for its blood.

There is, for example, the Philco Radio and Television Corporation, generated by the Philadelphia Storage Battery Company as a corporate life-saver in 1927 when radio was converted to use on ordinary 110-volt house current. Philco, we know, is a television licensee of Philo Farnsworth (the "young De Forest"). But Philco is also a licensee of RCA in radio. It is an extremely vigorous licensee, too. Nearly destroyed when general need for radio batteries was ended by technological advance, Philco has come back so strongly that it has sold more receiving sets than RCA in equal periods of time. In the good year of 1934, for example, Philco sold 1,250,000 out of 3,550,000 sets bought by the American public. RCA, which had given Philco its literal lease on life, sold a mere half million as runner up.[2]

By 1936 RCA was wondering, quite naturally, how on earth to restrain this galloping infant competitor it had loosed against itself. True, on every Philco set RCA received a royalty, but nothing relieved the strain upon RCA's own investment in manufacturing plant which was being assaulted bodily by loss of sales, or upon corporate pride. The obvious thing to do was to terminate the Philco license. But should that be done without some certainty of just what Philco was doing in television?

On July 30, 1936, Philco brought suit against RCA, the RCA Manufacturing Company, John S. Harley, Inc., Charles A. Hahne (or Hahn), and Laurence Kestler, Jr., charging them with unfair, wrongful, and illegal methods and practices, including the use of subterfuge, deception, false representations, and efforts to corrupt Philco employees and employees of Philadelphia Storage Battery Company by inciting them to breaches of trust and confidence reposed in them, in an "endeavor to entice, bribe, persuade and induce said employees to divulge and procure for them confidential information, data, designs and documents. . . ."

Hahne and Laurence Kestler, Jr., were accused of entering Philco's factory and therein and elsewhere putting themselves on good terms with numerous girls and young women in the employ of Philadelphia Storage Battery Company. This is in the spy tradition, but, reversing the tradition of spies among nations in which beautiful girls wheedle secrets from handsome young soldiers, the radio men took the

Philco girls over to see the bright lights of Philadelphia and then, according to the language of the complaint:

Did provide them from time to time with expensive and lavish entertainment at hotels, restaurants and night clubs . . . did provide them with intoxicating liquors, did seek to involve them in compromising situations, and thereupon and thereby did endeavor to entice, to bribe, persuade and induce said employees to furnish them for use by all the defendants, confidential information and confidential designs, all in breach of the duty of trust and confidence which said employees owed to the plaintiff herein and to said Philadelphia Storage Battery Company.[3]

This is only one of the more humanized passages of the Philco complaint asking the Supreme Court of the State of New York for relief from such actions. The others deal with more complex legalized aspects. And in Wilmington, scene of the old fight concerning De Forest and the cross-licensing agreements, RCA had to meet Philco on a second action seeking to restrain it from withdrawing the Philco license. RCA's publicity agent had just crowed, "our patents, which include the iconoscope and kinescope, have secured for the United States world supremacy in television." But in answering Philco's suit RCA's lawyers stated that, in seeking to extricate itself from agreements with Philco, RCA was only trying to forestall a nasty television patent problem. This would appear to be a contradictory state of affairs.

Philco is not to be exorcised by responses to lawsuits. It has just acquired license for a new television transmitter to operate with 15 watts of power in the 204-210 megacycle zone, and it still demands secure tenure of licenses and inviolate relationships with its employees. It still defends its trade secrets fiercely, and it continues to turn out more and more sound radio sets. And it keeps Philo Farnsworth snuggling closely, for all that it may have understandings with RCA and the Bell system. Philco must not be dismissed from the mind.

But whatever happens to Philco, there is still to be considered the matter of the Columbia Broadcasting System. The CBS is a program organization, pure and simple. Though it owns a few station licenses, in the matter of trade practices, it has an extremely high rating within the radio industry. With the public it is also relatively high in favor because of its "sustaining" features offered to fill in program time-space not sold to some commercial sponsor. The most famous of these has been for years the Sunday afternoon broadcast of concerts by the New York Philharmonic Symphony Society against which Mr. Sarnoff has only just lately countered with his NBC Symphony conducted by Arturo Toscanini.

Columbia is now installing a sound-sight transmitter in the Chrysler Building, to operate on a frequency in excess of 40 megacyles, with peak power output of 30 kilowatts. Its radius of reception will be about forty miles, and the definition exceedingly clear—about sixty frames of four hundred and forty-one lines each per second. It has retained Gilbert Seldes as program director, with a view to making television programs as good as those sound radio offers today. William S. Paley has announced that two million dollars would be spent to develop Columbia's technique of television broadcasting.

The trade magazine *Business Week*, wishing Columbia

well, pointed out that thanks to its control of basic patents, the Radio Corporation of America collects a license fee on every radio set manufactured in the United States. For, it pointed out, RCA could legally force the stoppage of the whole thriving set manufacturing business, if it wanted to, by refusing to renew licenses as they terminate.

The set makers entertain golden dreams of tomorrow's harvest when television becomes a commercial reality. But RCA is out to win the same dominant position in television that it holds in radio; and that disturbs the hopeful dealers.

The set manufacturers together with the broadcasting companies that entertain a similar concern about radio and television sending equipment, argue that a little competition might ease the situation; even two masters would be better than one.

It is because of these sentiments that the trade was so pleased last week with the Columbia Broadcasting System's announcement of its plans to install a powerful television transmitter atop the Chrysler tower.[4]

Evidently the enthusiastic seekers after competition were more eager than discerning. They should have inquired who manufactures Columbia's television equipment. And they should have inquired who is going to transmit the programs from station to station, for Columbia has no independence there. In the one case it must turn to RCA for radio relay equipment. In the other, the Bell system furnishes cables. And how do Columbia and RCA stand with the Bell system in the matter of using that transmission equipment?

The treaty of 1932 provides that RCA may use the Bell facilities for wire program transmission, picture transmission of material for programs, electrical sound recordings, one-way transmission of current for control of frequencies, and systems for radio program transmission or wire program transmission. On the other hand, the contract with the Columbia System provides that the facilities furnished by the telephone company are only for use in one-way radio program transmission. . . .[5] And so there goes our competitor, tangled in clauses and whereases worse than ever was Laocoön with the serpents.

What, then, has RCA to fear? It has fended off anti-trust suits and patent suits. It has its greatest competitor in set manufacture (Philco) on tenterhooks. It has Columbia, its great competitor in programs, buying RCA equipment and adversely placed in relation to RCA on the Bell transmission system. Mutual Broadcasting, the third largest program service in sound radio, has developed no known position of importance in television.

So, again, what has RCA to fear? There is always danger that someone in authority may hold that it is not serving the public interest, necessity, or convenience. And there is evidence of restiveness toward RCA. Representative W. D. McFarlane, of Texas, rose before Congress on July 19, 1937, and attacked the monopoly characteristics of the radio industry in particular; and on August 10 he made a second speech, going into detail concerning both Columbia and RCA.

It was exactly in this same way that the "radio trust" was attacked in 1929. When Mr. McFarlane spoke, the great Roosevelt boom was at its richest flower, just as the Hoover boom had been in 1929, and the radio industry, as before, only smiled as he demanded inquiry into its activities. But

before 1937 was ended another business depression had set
in as one had set in toward the end of 1929, and the radio
operators were becoming alarmed, if belatedly so, for they
are sufficiently skilled in public psychology to know that de-
pressions spur Congresses to "investigate." The statesmen
hope, somehow, by taking testimony and making findings,
to exorcise business miseries. And resolutions to investigate
radio were before House and Senate.

Mr. McFarlane's speech of August 10, 1937, "Radio
Monopoly Must Be Curbed," may be the unnoticed turning point in a new national policy concerning electronic
communications, or it may lead to nothing. We quote the
essentials of it here. They should be considered against the
background of facts the reader already knows as significant
indicators of what passes through the mind of the nontechnical radio critic in public office:

An analysis of the board of directors of the Radio Corporation of America bears witness to the correctness of the remarks of my colleague from Texas, Mr. Patman.

Gen. James G. Harbord is a Morgan representative on
the board of the Radio Corporation of America and is also
a director of the Morgan-controlled Bankers Trust Co. New-
ton D. Baker is legal adviser to many of the Morgan-controlled utility companies. Cornelius Bliss is a member of the
firm of Bliss Fabyan Co., a Wall Street firm, and is also a
director of the Morgan Bankers Trust Co. The elder Bliss
was for many years treasurer of the Republican National
Committee. Arther E. Braun, of Pittsburgh, is president of
the Mellon Farmers Depositors National Bank, one of
whose directors is A. M. Robertson, chairman of the Westinghouse Co. . . .

Bertram Cutler is listed in Poor's Register of Directors as

being connected with John D. Rockefeller interests. Edwin Harden, the brother-in-law of Frank Vanderlip, is a member of Weeks & Hardin, a Wall Street firm. Dewitt Millhauser is a partner in Speyer & Co., underwriters of utility issues. Frederick Strauss represents J. W. Seligman & Co., a Wall Street firm. James R. Sheffield is a corporation lawyer, a former president of the Union League Club and the National Republican Club. As a former Ambassador to Mexico he used his political connections with the Hoover-Coolidge State Department to get concessions for R.C.A. in South America.

Although the control of the Columbia Broadcasting System is supposedly a Paley family affair, the bankers are not without influence. When the Columbia network was purchased back from the Paramount Picture Co., the representatives of the financiers who put up the money for this purchase were added to the board. In return for the cash which the bankers put up they received approximately 50 percent of the Columbia Broadcasting System's class A stock. These banking interests were Brown Bros., Harriman & Co., W. E. Hutton & Co., and Lehman Bros. The members of the board of directors who represent these bankers are Prescott S. Bush, partner in Brown Bros.; Joseph A. M. Iglehart, partner in Hutton & Co.; and Dorsey Richardson, of Lehman Bros.

At this point I should like to say something about the Radio Trust formed by R.C.A., General Electric, Westinghouse, A.T.&T., et al., and which was supposedly dissolved by the Government in the notorious consent decree of 1932. Before the consent decree, R.C.A., who, under the illegal cross-licensing agreement with A.T.&T., et al., controlled the patents to radio-equipment manufacture, began to issue licenses to others—probably with the idea of convincing the Government and the public that they were not such a bad trust. But, after the consent decree, I have learned of no licenses for radio-set manufacture that were given by R.C.A.

When the Government seemed to be pressing suit against the Radio Trust, the cost of radio sets dropped to $10 and below. This permitted millions of homes to enjoy the benefits' of radio, and millions of people were able to listen to the issues of the day aired over the wave lengths. A new note in democracy was being struck. However, just as soon as the Hoover administration and the Radio Trust entered into the now infamous consent decree the price of radios began to rise again until now $30 and up is the price for a decent radio.

Not content with their monopolistic control and the 5 percent on gross revenue they take from all licensees they began to terrorize even those who had licenses to compete with them. The case of Philco Radio & Television Co., which filed a suit against the R.C.A., charging espionage and other terroristic practices to R.C.A. is eloquent testimony.

Other independents, if they desired to compete, were forced to run the gamut of patent-infringement suits brought by R.C.A. To fight a case of this sort costs a great deal of money. The adjudication of a patent through the Supreme Court sometimes costs over $100,000. Such a cost is prohibitive to most independents. His choice is due in one of two directions: Either he fights and the cost of litigation plus threats to his customers drives him out of business; or, he wisely goes out of business upon the receipt of a threat of an infringement suit. In either case, the independent gives up the ghost. Such is the power of the patent racketeering of the Radio Trust. . . .

In the supposed dissolution of the Radio Trust by the consent decree in 1932, it was proven that R.C.A. possessed such a monopoly. There is evidence to show that despite the consent decree, this monopoly still persists in violation of the anti-trust laws. Yet testimony before the House Appropriations Committee shows that broadcasting licenses of R.C.A. are renewed every 6 months without ever having the question of the apparatus monopoly or public interest raised. I sincerely believe that the issue of reëxamining the effects of the consent decree is resting squarely on the shoulders of Congress. Shall we face the issue or evade it as has been the custom in the past?

The gentleman from Massachusetts [Mr. Wigglesworth], who, as a member of the Appropriations Committee, has given much time and consideration to this subject, has spoken several times favoring the immediate clearing up of this communications monopoly. His work in the committee bringing out the existing known facts, I am sure, has the hearty approval of the Congress. Several other Members have spoken, pointing out the great need of an investigation. . . .

Mr. Voorhis. Mr. Speaker, will the gentleman yield?

Mr. McFarlane. Yes; I yield.

Mr. Voorhis. Does not the gentleman feel that perhaps the root of this whole matter is to be found in the fact that these corporations have been able to call a certain radio channel their own; that, as a matter of fact, if there is any natural resource that ought to belong to the people it is the air, and that we are gradually building up here a vested interest in the ownership of channels of communication through the years? Would not the gentleman favor some tax measure which would levy a good stiff franchise tax and take the water out of the situation so that the only advantage would be a temporary license, or a license running for a certain period of time? Would not this prevent the building up of a vested interest in these channels?

Mr. McFarlane. Answering the gentleman, I may say that there has been tax legislation pending before the Ways and Means Committee since the early part of this year, but we have been unable to get any action on it. This would require the radio industry, which is the only public utility operating in interstate commerce in the United States today that does not at least pay the cost of its supervision, to pay a suitable tax; but this bill, like the others which should have been brought to the floor of the House, never has been considered by this committee and still lies buried there.

Mr. Leavy. Mr. Speaker, will the gentleman yield?

Mr. McFarlane. Yes; I yield.

Mr. Leavy. The gentleman's remarks indicate that he has given much thought and study to this question, and he is making a strong case. I am wondering if he has covered the further abuse that is generally recognized of large, metropolitan newspapers of the country acquiring radio stations and then hooking in with the great radio chains and thus controlling channels of news through radio as well as through the press?

Mr. McFarlane. If the gentleman will read my remarks of July 19, he will see that I dwelt upon that very question; that I pointed out that some 200 of the large daily newspapers of this country own the largest radio stations in America, and through this method of radio broadcasting and sound motion-picture equipment and through the press, through that tie-up, they absolutely control and mold public opinion in this country today; and this is why Congress is having such a terrific fight to get any worth-while legislation enacted for the benefit of the people. [Applause.]

Mr. Wearin. Mr. Speaker, will the gentleman yield?

Mr. McFarlane. Yes; I yield.

Mr. Wearin. That tendency on the part of the newspapers coupled with the operation of the present chain does constitute a serious threat in the way of a monopoly to influence public opinion, does it not?

Mr. McFarlane. There is no doubt about it.

Mr. Wearin. I am sure the gentleman is familiar with the fact that I have a bill now pending before the Committee on Interstate and Foreign Commerce to prevent a continuation of this monopoly.

Mr. McFarlane. I know the gentleman has had such a bill pending for some time, but he does not seem to be able to get action on that any more than the rest of us are on these other bills. We cannot, apparently, get these bills out of these committees which would be of such tremendous benefit to the people. And this communications monopoly is becoming more powerful all the time. Until now many Members dare not speak their sentiments against it, lest they be opposed by it for reëlection. . . .

But does the grasping of the monopoly stop there? Let me quote the following from the Hollywood Reporter of July 1937:

R.C.A. Now Believed Aiming to Control Communications

Washington.—There is a well-authenticated report that the Department of Justice is now willing to withdraw its objections to a merger of Western Union and Postal Telegraph. In inside circles this is seen as an indication that R.C.A. is moving to control the entire communications field.

The ultimate battle, of course, will come over the control of commercial television. In view of President Roosevelt's determination for a unified communications system, it is possible that if the big wire companies merge, R.C.A. might let the merged outfit have the communications business and devote itself to the amusement field and broader television activities. However, this possibility is not credited by those in the know.

They believe that R.C.A. will make every effort to control both Western Union and Postal in an effort to broaden its telegraph business, and that the fight will then be between R.C.A. and A.T.&T. for full control of both communications and television.

It is not thought possible that if the wire companies do merge, the new company would be able to protect itself against the threat of radio competition by acquiring R.C.A., the supremacy of R.C.A. being seen as much more logical. In any event, it is believed that the merger would make commercial television much more imminent.

There is an interesting sidelight to the relation between R.C.A., G.E., and Westinghouse, but nevertheless important, and bears mentioning here.

When General Electric, Westinghouse, and R.C.A. were busy dividing up the radio field amongst themselves a very peculiar transaction took place. In return for certain stock and physical assets given to R.C.A. and which R.C.A. itself valued at $42,864,812 plus the exclusive manufacturing rights and the royalties to radio device field, General Electric and Westinghouse received 6,580,375 shares of R.C.A. stock, the market value of which was $263,215,000. In other words, R.C.A. paid $220,350,147.50 for the exclusive rights in the radio-apparatus field, and gave the control of R.C.A. to G.E. and Westinghouse. The facts are borne out in an unchallenged affidavit on file in the Federal court. It is difficult to believe that they were worth that much. It is far easier to imagine the innocent investing public who owned R.C.A. stock, through no choice of their own, made a gift of these hundreds of millions of dollars to Westinghouse and General Electric. And from the message I read to you earlier from S.E.C. the law is unable to cope with this manifest racketeering.

I want to ask that the committee now investigating tax evasions and tax loopholes investigate this gift of $220,000,-000 to Westinghouse and General Electric and learn just what taxes were paid on this $220,000,000. I also ask that they report their findings to this body.

The people of the United States have paid $2,262,375 last year to regulate the communications industry. In all other kinds of industries operating under Government franchise the cost of their regulation is placed on the industry. Why, then, should the taxpayers continue to keep up the cost of the Federal Communications Commission? I think it is now time for Congress to shift this burden from the shoulders of the taxpayer on to the communications industry, which operates under Government franchise for which they pay nothing.

I cannot repeat too often the query, "What does Congress intend to do?"

It is a wise monopolist who knows when to ease up. RCA may not be the sort of institution that Mr. McFarlane and his colleagues think it is. All these things may have been said in misunderstanding of the facts, but there the facts stand and the opinions with them. The stockholders of RCA, battered as they are by the years of tribulation and lawsuits, have the record to ponder. The public, who may be called on to finance the development of television either through direct governmental subsidy or by purchase of equipment at original high prices, have some things to ponder, too.

They may think upon the reference in Mr. McFarlane's speech to the prices paid for receiving sets during and after governmental anti-trust actions, and upon the fact that RCA, like every other radio operator, must depend finally upon the "public interest, convenience, or necessity" for its license to exist in the broadcasting industry. RCA could continue to make equipment if barred from interest in stations, but it wouldn't be happy under such circumstances.

20. Return of a Pioneer

TELEVISION IS KNOWN WITHIN THE TRADE AS A "LOCK AND key" business. Transmission and reception are bound together in a mechanically monopolistic way, no matter what the courts or commissions say about legal monopoly or its absence; and there seems no way to extricate them from their relationship.

Of course, sound radio is a lock and key business, too, to a certain extent. Without an adequate receiver in operation, the broadcast is futile. But in sound radio, selectivity of programs is not very difficult within the framework of standards now developed. The average receiving set can tune in on from ten to a hundred broadcasting stations. Its dial spins with the world. But that cannot be the case with television as we know it according to present engineering development.

But Mr. Jewett, Mr. Espenschied, Mr. Sarnoff and Dr. Jolliffe have made it emphatically clear that the spectrum does not accommodate television broadcasts in either number or range comparable with sound radio. Today, the average city is served with three to seven sound broadcasting stations; but tomorrow these will be gone, and only one, two, or in the rarest of instances, three television programs will be available.

In the second place, the technical nature of television does not allow any variations in equipment comparable with those of sound broadcasting. We now have inexpensive little radios for bedroom tables, "high-fidelity" console types for the drawing room, and special kinds for automobiles. They vary in sound definition without losing entirely their ability to compete with one another in the actual reception of the radio signal.

Not so with television. If a program is scanned at the rate of four hundred and forty-one lines, sixty frames per second with RCA's iconoscope, then no set can translate the electronic impulses back into comprehensible pictures except one designed especially for reception of a four hundred and forty-one line, sixty frame iconoscopic broadcast. And so it goes. If scanning is done by use of pierced disks, helical arrangements of mirrors, or any other variation of mechanical systems, then only receivers geared to these scanners can function.

It is immediately apparent that some basic standards must be set: television must be all of one thing or another, technically speaking, if it is to arrive commercially. And so, when we consider all that has occurred in sound radio, we recognize the enormous responsibility placed upon the group which sets those standards. These technical qualities will, in the end, resolve all questions of television competition; and monopoly by exclusive patent holders will give them dominating positions. There are two trade associations in the radio industry which speak for all competitors in general in the resolving of these pressing questions, just as counsel and legislative friends speak for interests in particular. These two trade bodies are the National Association of Broadcasters and the Radio Manufacturers' Association.

NAB is the spokesman for the disseminators of programs. It encompasses more than sixty-five per cent of all station operators, and these do in excess of eighty per cent of all advertising business in radio. At the F.C.C.'s engineering conference in 1936 James W. Baldwin, managing director of NAB, suggested on behalf of his organization a plan of allocation which would provide eight television channels below 100 megacycles, but pointed out that they would not be enough, really, for the demand.

There are, however, more than technical considerations involved here. The American broadcasting system is a competitive system. It is a great system because it has been competitive. . . . [A relative term, you will recognize.]

And our plea is today that you allow television to develop on the same basis. Better we delay the introduction of television than in enthusiastic haste inaugurate it and find that through control of patents so powerful an instrument is in the hands of too few people.

If television were ready to be inaugurated on a basis of a national competitive service, he argued, then the F.C.C. was clearly under a very great responsibility in determining in advance whether for all practical purposes the ownership of basic patents and agreements, if any, between patentees would permit competition in the construction of television transmitters and receiving sets. He seems to have an unassailable position there.

We should also know in advance what relationship, if any, may be established between the sending and receiving apparatus. Will there be freedom in the selection of receiving sets or will the use of terminal facilities be controlled in a manner comparable with the telephone?

Surely everyone will agree that those who own television patents are entitled to a rich reward for their creative work, but because of the public service inherent in television, patentees should be denied the right to control its use. Keep it from the hands of monopoly and allow it to develop only on a national competitive basis.[1]

But just how valid, in view of the facts, is the chance of competition? Technically, television is a lock and key operation. Ownership of lock and key is decided on the basis of a patent position. A patent position, we will all admit, can be developed more easily by the rich and politically powerful than the poor and weak. And once a patent position is attained, the holder of a patent has the right, under the Constitution of the United States and the findings of the Supreme Court, to make or not to make the article patented, to lease or not to lease rights to others.

What Mr. Baldwin asks for, essentially, is an "open patent pool," of the sort ordered in Great Britain when television became a public institution in 1935. Such a group must allow anybody to join it who has a patent of value to contribute. And the contributor thereupon has common power with all the other participants to use any combination of the pooled patents he so desires to make a set of his own design, paying royalties to the particular contributors whose patents he happens to use.

If such a patent pool could be arranged in the United States, who would participate? Already, the sound radio industry is dominated by two basic organizations, the A. T. & T. and RCA. Could there be true competition so long as these two major operators continue to follow the lines of policy indicated in the treaty of 1926, as amended in 1932? Hardly.

But what about the others who attended the engineering hearing with Mr. Baldwin? Said the Radio Manufacturers' Association spokesman, James M. Skinner, who happened also to be the president of RCA's troublesome licensee, Philco:

RMA has tried to crystallize the basic needs of television in a five point plan:

1. One single set of television standards for the United States so that all receivers can receive the signals of all transmitters within range.

2. A high definition picture approaching ultimately the definition obtainable in home movies.

3. A service giving as near nationwide coverage as possible.

4. A selection of programs, that is, simultaneous broadcasting of more than one television program in as many localities as possible.

5. The lowest possible receiver cost and the easiest possible tuning, both of which are best achieved by allocating for television as nearly a continuous band in the radio spectrum as possible.[2]

We are thoroughly familiar, by now, with Mr. Skinner's problems and objectives. One set of television standards—we know this is essential for uniform reception of the sort now common in sound radio. We also know it entails inevitable monopoly. A high definition picture—this is simply a test of consumer interest. Unless the picture is large, clear, easy on the eyes, television naturally could have no interest

for the ordinary person. And as to universality of acceptance, the more nearly possible it is to distribute a single
program across the country the more nearly will advertisers, if that type of exploitation remains, be able to achieve
the highest possible consumer interest. Also, the more
nearly will a political candidate be able to reach all the people simultaneously. And the more nearly will all the people
be able to see, as it happens, some major news event.

And here we stumble again upon the difficulties of program selection. Mr. Skinner reminds us of the clash between Messrs. Jolliffe and Jewett, when he points out:

> It must be assumed that if a given channel is assigned in
> Boston, that channel cannot be assigned to any other cen-
> ter nearer than Philadelphia, and any channel assigned in
> New York cannot be assigned again any nearer than Balti-
> more or in Buffalo. Similarly, any channel assigned in Cleve-
> land probably cannot be assigned in Toledo, Akron, Youngs-
> town, Buffalo or Detroit.
> It is not likely, at least in the early days of broadcasting,
> that adjacent television channels could be assigned in the
> same city, because of probable interference.[3]

Nationwide service must be the goal of television, Mr.
Skinner feels, even though that really results in exclusive
operations in a given locality by one licensee. This condition we know must often be the case, for sets designed to
receive programs broadcast on one frequency and definition
cannot make coherent the programs sent out on another.
Here he falls into a contradiction.

Mr. Skinner admits that it will be difficult enough even
to distribute a single program on a nationwide basis, but in-
sists, nevertheless, that competition and the public interest

be served by offering the residents of a single community at least two television programs from which to choose. How to resolve this conflict with his principle of imperative nationwide service, Mr. Skinner does not say. The public must naturally bear the cost of distribution, the RMA feels. The history of cost is enlightening. Sound radio sets, between 1924 and 1929, cost on an average of one hundred and ninety dollars, only to drop in price to present levels after technology (and lawsuits) had provided higher standards of efficiency and lower costs of operation through experimentation.

Finally, Mr. Skinner states:

In the opinion of RMA, the Federal Communications Commission has in television a great opportunity and a great responsibility. Here is an impartial body, and with no interest to serve but the public interest.

The public is already aware of television. The public not only wants television, but it expects television, and it seems to be getting somewhat impatient over the long time it is taking to work out.[4]

All these are valid words, and significant. They convey as much of warning as of invitation to the Communications Commission. The public and Mr. Skinner are sitting in to see that the commission does not forget that it has no interest to serve but the public interest.

Which brings us to an examination of the commission. Let us re-emphasize the importance of this group's position. It must set standards of performance which will have infinitely ramified effects. Here are just two examples of the repercussions which may be expected from its decision:

The television facsimile service may lead ultimately to a decision on whether radio or newspaper interests will control dissemination of news. . . .

Facsimile will broadcast a full newspaper, banish newsboys, presses, delivery systems, make dot and dash telegraph as obsolete as the pony express, by visual transmission of information, weather maps . . .[5]

Labor recognizes the threat inherent in such an event. William Green, president of the American Federation of Labor, observes "radio is more important to the public welfare than the newspapers."

And in the second example:

Television appears to be a rich and fluid medium, and writers and directors, especially, might be eager to see what they could do with it.

At this point, however, some cold realities of engineering and economics intrude themselves. A television channel is an exceedingly costly thing, running into hundreds of thousands.

Further, the great plaint that radio uses up literary material too fast (one broadcast on one evening and the manuscript is finished for all time) is as nothing compared to what television will do to stage settings.

A theatrical producer, planning a season's run for his play, can invest in substantial settings, but what would happen if he had to change his play and his sets not only every night but several times in a single night? Again, the cost approaches the fantastic. . . .

In this dilemma, the Farnsworth studios are working on an ingenious solution based on the use of miniature sets.[6]

But suppose all these problems of frequency allocation and program detail are settled. We cannot resist the indication of another, a problem supposedly settled at the outset, concerning technical quality of operation itself. This is a bit of news from a laboratory most people believe quiescent. Nothing could more sharply and dramatically remind us of the ceaseless dilemma with which the regulatory commission is confronted, as it seeks simultaneously to protect and foster public interest and private enterprise, than:

I would like at this time to advise the Federal Communications Commission that we have designed and patented a mechanical system [of television operation] in which a new and revolutionary principle is involved.

The principle is so radically different from that of any other system heretofore used that it would not be possible to adapt any of the present existing methods of inter-laced scanning to this system, although it does utilize inter-laced scanning.

This receiver is capable of projecting a three foot square picture with a definition of two million picture elements, and although this definition is considerably higher than that contemplated by some of the present companies, it has been demonstrated as commercially practicable. . . .[7]

By now the reader realizes how sensational a statement this can be, if true. Out of the tortures and tribulations of sound radio, RCA and the Bell system have laboriously built themselves to powerful positions in communications. They have won after battle and compromise. Great sums have been paid out in lawyers' fees and other costs to develop patent positions giving them dominance. And dominance, insofar as RCA is concerned, is predicated upon the cathode ray scanner, the iconoscope. And though the Bell system is not promoting any particular type of scanner, it is into this field deep enough to protect its stake in the coaxial cable, which it hopes to force RCA into using, in preference to the relay, point-to-point booster of radio waves.

Who challenges the giants, then? It is R. D. LaMert. And who is he? R. D. LaMert speaks for the De Forest Television Company, of Hollywood, California. This is the truest sort of drama. Lee De Forest, who is called the "father of radio," the great elder, the inventor who got a pittance and could not keep it, returns to the wars with a new invention, threatening to control the new art.

De Forest can say, without any fear of challenge, that RCA and the Bell system are nothing more than the corporate expressions of his own genius. By inventing the three element thermionic valve, he made them. By inventing the radical mechanical television scanner, will he unmake them?

This question can be answered in one of two historical ways. De Forest is old, and he is almost alone. He can fight at this late day through the Supreme Court of the United States, but is unlikely to do so against these two aggregations of power. He can surrender and sell his wonderful new device to one of the two survivors from the early battles. He comes close to holding a balance of power over his ancient enemies and partners. The Supreme Court may yet decide between him and them.

22. Public Policy

AND SO WE COME TO THE CONCLUDING QUESTION, "WHAT shall we do about television?" There can be no challenge to the use of "we," for the declared policy of the Republic is that interstate commerce in electrical communication, whether wire or wireless, shall proceed only in the public interest, necessity, or convenience. We have something of an understanding of the basic principles upon which radio technology is founded. We know the history of custom and law developed with the changes in the technology.

There can be no challenge to the statement that the future of politics and social order and the future of television will follow parallel courses. Sound radio has already precipitated the fall and rise of governments. On one occasion (March 4, 1933) it stabilized a great nation gone hysterical. Nobody can recall the first inaugural address of President Franklin Delano Roosevelt ("The only thing we have to fear is fear itself") and deny the social importance of radio communication.

But what is the status of administration? Some portions of the geographic United States do not receive enough service; others receive too much. Sound broadcasting is contaminated to the extent that it is almost always thought of by the common listener as an adjunct of commercial advertising. The financial organizations most closely connected with broadcasting work toward monopoly and suppress technology in the interests of business stability. Programs are offered not always with the highest motives, not chiefly for entertainment and instruction, but primarily for the purpose of propagating sales of goods. Licenses fall too many times into the wrong hands. These are commonly uttered criticisms of the sound radio broadcasting industry. They should, however, be directed not against the commercial interests but exclusively to the Federal Communications Commission. For the commission is the people's representative in a convergence of operators who have been authorized to function solely in the public interest, necessity, and convenience. Yet the people's representative has failed to declare positively just what the public interest, necessity, and convenience encompass.

If there are too many radio stations in some sections of the country and not enough in another, has not the commission power to remedy? Obviously. The records of its proceedings are filled with complaints by radio interests that the "spectrum is too crowded." Yet the number of stations increases. The commission revokes only about two operating licenses a year, and it allows the net total to increase. Yet it does not distribute individual stations to the best geographic advantage.

If the operators devote themselves more to selling time-space for advertising and less to good entertainment and education which bring in no cash returns, has not the commission power to revoke licenses in cases of neglect to the public interest? And can it not do the same when political

127

utterances guaranteed under the Bill of Rights are censored? Can it not do the same when the canons of decency and good manners are violated? Can it not do the same when any matter of public interest, pertaining either to the quality of goods advertised or the nature of news events, is misrepresented or ignored? We know that it can do all of these things, for it has absolute powers of determination as to what is encompassed by the term, "public interest, necessity, or convenience."

And we know that if it does not do its duty, the fault is our own. It is illogical and unfair to expect the Radio Corporation of America meekly to surrender its monopolistic ambitions, or the Bell system to write off as a financial loss its great network of telephone wires in the face of technological change. It is absurd to think that either of these institutions, if it can buy up the patents and devices of some inventor like Philo Farnsworth or Lee De Forest, will not use these to its own profit, regardless of the effect upon the public or upon competing business groups.

Sound radio has demonstrated that the communications business is conducted on the basis of a titanic struggle to protect huge investments and attain great profits. It has demonstrated that the very essence of use in a communications instrument demands an attempt by the exploiter to monopolize the field of operations. The more nearly universal the acceptance of a device is, the more valuable it becomes not merely to the exploiter but to the user. A telephone that connects the international business man with Paris, Bangkok, or Berlin is more important to him than one reaching only into the next county.

We have not dealt in detail with many criticisms of the

Communications Commission which, however valid, are relatively minor. Analysts of administrative government hold that it has a faulty structure. A seven-man body charged with both executive and judicial functions tends to become a debating society, they hold. A similar statement might be made about the Supreme Court of the United States, unquestionably the final source of power in our constitutional government. The tendency to debate rather than execute duties would seem to depend upon individual will rather than mere communal session.

The commission has been made the butt of political manipulation and patronage abuses. This is one of the most common criticisms of all, but it is also one of the weakest. Political manipulation and patronage abuse can be carried on under any form of government; and so they are. In a democratic republic they proceed exclusively by sufferance of the citizenry. If the public objects to maladministration, it knows the remedy it has itself provided.

There are some criticisms of the communications law which must be considered, too. Investors in radio financing and operators of radio systems declare that the licensing provisions leave them in a state of nervous apprehension. A man has no tenure, they say, no certainty that he will be allowed to continue operations for more than six months. They ask licenses for not less than five years at a time, and also some guarantee that property rights will be protected in the event renewal is denied. The history of administration offers no tangible basis for such fears. Licenses generally are revoked only in the most flagrant instances of abuse or financial inability to perform. None of the great broadcasting chains or their stations has suffered such punishment yet.* License holders, except in rare instances of experimental operations, are not even required to carry the burden of proof that they serve the public interest when they come in for renewal of co-operating permits.

A five-year license obviously would insure the operator more adequately against the perils of technological change; but it would also operate to restrain advance in public service.

There is a strong congressional movement for limiting the licensing rights of the commission in such a way as to prohibit joint ownership of radio stations and newspapers. The theory appears to be that radio and newspaper should compete for the news, and offer contrasting editorial opinions upon the events of the day. Such a condition, it is held, would insure against monopoly of information and distortion of the public mind. Here, again, the remedy lies not in legislation but in public action. If the law allows development of great chains of commercial broadcasting stations of the sort typified by National Broadcasting Company, and continues the present practice of granting broadcasting licenses to set up facsimile systems of broadcasting, the printing press newspaper may find itself no longer an important instrument of competition. It, like the legitimate theater in relation to motion pictures, tends to become just a sort of testing service.

Radio is being monopolized at the source, not at the outlet. The final licensee operator of a station, whoever he may

* The only indication of a change in policy toward these was made by Chairman McNinch after the commission held that NBC had violated the canons of decency in the Mae West broadcast. He said that case would be considered when the fifty-nine stations using the program in question applied for license renewals thereafter.

be, is at the mercy of the Federal Government and the great organizations typified by the Bell system and RCA. His dilemma is in no way solved simply by making him a radio operator-groceryman, rather than a radio operator-publisher. And the newspaper publisher, presumably skilled in the difficult art of satisfying a great majority of the community as to entertainment and information, possibly is more to be trusted than the grocer, the banker, or the insurance executive in developing good radio program policy. No evidence has been offered that publishers have been or are worse or better than the average station licensees. They just go into radio as rapidly as possible, simply as a hedge against the day when facsimile may put the press out of business.

The spirit of this proposed limitation upon them might better be preserved by granting licenses only to bona fide residents of communities in which stations are operated, and upon proof that the public is being served in the best manner possible. By this means none of the values in nationwide broadcasting of single programs would be lost, for chain networks would continue but local interests would be more likely to dominate in editorial handling of news and politics.

Inescapably, as one ponders these problems raised by sound radio and shadowing the future of television, one finds there are three basic questions:

First, shall radio, sight and sound, continue as unrestrained, untaxed, private enterprise under the present system of licensing?

Second, shall it become a closely regulated public utility,

with fixed rates and tariffs comparable to the telephone industry?

Third, shall it be liquidated as a private enterprise and operated exclusively by the Government?

Let us keep in mind the historical background as we examine these three, seeking to analyze the future of television.

Television, structurally, is a synthesis of communicative forms. It combines sight and sound. Operatively, it is as the lock and key. The whole function of the manufacturer is to serve the holders of the lock and the holders of the key, the transmitters and the receivers. Television is a medium of information and entertainment for the control of which a terrific struggle is being waged. It is a medium also for acquiring great profits both in money and in power. It is coming into ordinary use slowly; but if the history of invention is not to be denied, television will in time become as common as the sound radio is today. It is expensive now, but ultimately the price will meet the market demand because technological advance and change have been found to achieve such results, however incidentally.

Technology, in itself, guarantees nothing save change. When inventors have succeeded in developing adequate standards of performance they have done their job in life. It is no duty of theirs to be concerned with bankruptcy courts, frenzied investors, price structures, vested interests, or the trading philosophy of the Bell systems. When De Forest discovered that a grid would modulate the flow of electrons, he had merely to put the grid in its proper place; and so it was with Edison, Hertz, Marconi, Alexanderson, Steinmetz. How easy, compared to the plight of the business man, the lawyer, the stockholder, and all those who would regulate the economy in which television must find a place!

These inventions, in overcoming the problems of techniques, create engines that strain and sometimes destroy the economic structure expected to accommodate them. And the imperative of accommodation is one the lawyers, the business men, the economists and commissioners, try as they will, cannot deny. A telephone is invented, and soon all must have it; and so too with the electric light and the automobile. Government fiat, suppression by vested interest, and the activity of those who stand to lose by the development of an invention may delay and harass its progress, but if historical precedent is to have any meaning, we have to admit the invention is accepted finally. Unfortunately for our peace of mind the familiar institutions of profit, free private enterprise, free price, private property, which have regulated our economy are inadequate to effect a painless acceptance of the new state of affairs when invoked in their pure forms. They were developed in a handicraft economy, and we find them unable to function freely in conjunction with the highly mechanized, integrated machine technique. But after the clash of inventions and established institutions, modifications in the character of control always seem to take place, for an invention is used as it is conceived—or it is not used. That is, a machine is a machine and nothing more. It performs one task and that inflexibly. It is control that must be flexible and adaptable. And flexible control has always managed to make a place for new machines so far. How will adjustment come about in the case of television?

Upon the resolution of our three basic questions rests the fate of industries with investments in the billions of dollars: the future of communications, the test of government regulation, the radio industry's subdivisions of transmission, manufacturing, and advertising; the motion picture industry with its technicians, furriers, heroes, and heroines; the telephone monopoly and its seven hundred thousand-odd stockholders and all those dependent upon them.

Whoever gains the initial advantage of pre-emption will have a major power over the many others who, in order to continue existence, must have a part in television. But being first has its perils; and this is a warning to sound radio in its fight for control. The initial investor must cope with the capriciousness of technological change; the economic yardsticks of profit and loss; the Federal Communications Commission's amorphous definition of public interest, convenience, and necessity; and a mystical winnowing of all these in the flailing chamber of the United States Supreme Court.

Will the first entrepreneur in television serve in the manner of the male bee, simply to fructify and die? There is strong chance of this, for television requires heavy investment for plant, personnel and operating material. Errors in judgment, therefore, will be penalized severely. Uncertain, faltering regulation will be fatal to all concerned; industry, the public, the Government, the economy in general. This ought to be obvious, but seems not to be. Unquestionably capital stands ready to bring out television. But upon investment there must be a return. If television is to be privately operated, those interested must recognize that regulation is a technical imperative, and that the healthy condition of the art requires regulation to be stringent and honest. For, to attain a measure of the precious stability necessary for successful commercial operation, the interested parties within the industry must know on more than a special privilege basis where they stand in regard to regulatory administration and policy.

Special privilege is an ephemeral thing. What can be achieved today may tomorrow be passed on to a richer, more influential interest. Operating on a six months' license, or even a three or five year license, the entrepreneurs are entitled to know what definition is going to be applied to that much too mysterious phrase, "public interest, convenience, or necessity." By the same token, they must consider seriously the elevation of program standards, the problem of balancing between "editorial selection" and censorship. An aroused public opinion is at times very costly to investment. If the public has been lax to the implications and operations of aural radio, that does not mean it will be lax with television. The growth of "Legions of Decency" and the rumblings in Congress are indicative.

The Federal Communications Commission will be a potent factor for good or evil, of course. Its history in the radio broadcast fields is well known. What will be its position in television? It will decide the problems of allocation on the spectrum, fix standards and interpret the institution of law as it relates to the art. The question of listener and viewer interest will be posed to it. What will be the rights of the receiver of programs if the construction of a steel building in his vicinity interferes with technical reception? What if a doctor's diathermy machine conflicts with a program? But most pressing and immediate are the definition of the public interest, convenience, and necessity as applied to the granting of licenses, the allocations on the spectrum, and the fixing of standards.

In addition to the spectrum problems, there are standards of performance to be fixed. Transmission and reception are reciprocally dependent. One type of transmission technique requires a similar type of reception technique. And there are many inventors, each claiming that his instrument should set the standards.

But the problems of monopoly, place on the spectrum, and technical standards are soluble by simple fiat, or economic strength. They are just empirical tests of quality. They are not imponderables, merely problems involving exercise of choice. But there is another problem here that tortures the sleep of the business man: once television is out, can it be made to pay profits and remain stable in technical development? There, somewhat simplified, is the crux of the present impasse. Adapting the electron to create pictures is one thing; making it profitable is another. Some pretty problems have resulted from the attempt to squeeze television within the institutional framework of sound radio which operates on the principle of selling sets to the consumer and charging the cost of programs to the advertisers. In television the consumer can still buy his set, if the price is right. Some have attempted to pose this as a focal issue. Actually it is not, for under the system of mass production the price of sets undoubtedly can be brought down to reasonable levels. It is a good risk to say the audience is ready.

But the cost of programs is really maddening to those who would share in the television harvest. To keep it within the framework of sound radio this cost should be borne by the advertiser. Can he do his part? Experience has demonstrated that the cost of programs has been reasonable in relation to advertising potentials. This system apparently has been profitable to radio men and advertisers alike even at the price of $20,000 for one hour's entertainment on a national chain program. This is what Mr. Sarnoff and Mr. Paley call the "American Way."

And in television they would also like to operate in the "American Way," of course, but television is very unpatriotic and up to the present it doesn't seem able to conform. For one thing, as has been noted, its character is embarrassingly monopolistic, setting it counter to traditional competition. Yet more embarrassing is the search for someone to bear the cost of programs, most of which, it is well established, will be in the form of film motion pictures. A motion picture feature giving an hour's entertainment costs from $350,000 to $1,000,000 and sometimes more. What advertiser can bear this cost for an hour of television?

But should it be found that the cost of film can be circumvented or solved, another difficulty arises. Will the eye accept advertising in television? The experience of motion pictures has demonstrated that attempts in this direction are dismal failures. In 1937, for instance, in a city in Missouri, groups of "movie" patrons took it upon themselves to boo and shout catcalls when advertising appeared on the screen. This appears a very bad omen—but do not underestimate the advertisers and radio men. For instance, it has already been found that by reducing the size of the film from the standard thirty-five millimeters to sixteen, substantial savings result. By the use of miniature sets in perspective against neutral backgrounds, more economy is accomplished. Those vested with the guidance of radio and advertising are resourceful and ingenious. And they are working frantically to solve their difficulties.

One fact remains: however much they seek to work separately, still they must come together. Progress in engineering is and always has been a cumulative expression of all technical information. The engineers of RCA do not work isolated from the world any more than their competitors do. They exchange information, rush to their laboratories, try to accomplish new results ahead of the other fellow. So do stage designers, managers of performance, financiers, and program directors. Then, the new analysis supposedly achieved, the triumphant one demands of his government protection in the form of patent, copyright, judgment for damages in plagiarism. He seeks external help because alone he is helpless.

Clearly, then, television cannot escape government dominance, however much effort is expended to make it conform to the principle of free, private enterprise. It is simply a matter of how much dominance. There are business men who want to remove television entirely from the influence of the traditional sound radio technique. They are especially concerned about the effects of the synthesis upon motion picture exhibitors and their interests. Why not set up television as a public utility? Thus do we find ourselves sliding from the premise of free competition and free profits to limited and guaranteed income in return for guaranteed public service on a common carrier basis, comparable to telephony.

This, it is claimed, would make possible more effective

regulation, and permit the economic stability in television so notably lacking in sound radio. In addition, the problem of cost, now so difficult for those who would like the competitive advertising method, might be solved simply by charging service costs to the consumer on a utility rate basis. This proposal has been put most concisely by Robert Robins, executive secretary of the Society for the Protection of the Motion Picture Theatre, an organization of independent theater owners, radio set manufacturers, and other imperiled interests. Mr. Robins, appearing before the Informal Engineering Conference of the Federal Communications Commission, outlined a three point program.

Television service in its early stages, he held, must be confined to entertainment and educational purposes, such as the regular motion picture feature production, shorts, and newsreels; and television must be kept free of advertising. Furthermore, the programs must be a separate and distinct service, must be offered to the recipients on a service charge basis, and rates, rules, and regulations must be determined by a competent public body.[1]

His plan is very persuasive. The cost element is solved if consumers will pay. The difficulty here is that the possible consumers of television programs are the same people who now own radio sets. Their habits of thought have been so conditioned to receiving what appears on the surface as a free service paid for by the advertiser that the success of an attempt to burden the public with program cost directly is at least questionable. On the other hand, payment for electric light and telephone service is generally made without complaint. Television could be added to either of these without undue bother.

But other difficulties appear. If the transmission from studio to the home is to be brought through the means of the spectrum, the frequencies upon which television operates cannot be staked out or fused in to permit only qualified rate payers to enjoy their benefits. Individuals deserving to receive television programs but not anxious to pay the required fee will be tempted to buy or build receiving sets and "bootleg" programs into their homes as they have in England.

Strict competition would be absent from such a scheme. To allow television to develop on this basis in the hands of two or three public service companies suggests that those paying a fee to one transmitting company will be unable to receive the programs of other transmitting companies. If, to overcome this situation, the fees and rates are equally divided among the transmitting companies and the service of all broadcasters is open to all consumers, the incentive to supply better programs in order to attract more listeners is dulled. One just shares the current income and lets new business come on when it wills.

Why not throw the private interests out the window and put the whole matter in the hands of the Federal Government? The Government, it may be contended, could maintain the most elaborate sort of programs, extend the technical operating facilities over the widest physical areas, and continue development of the art and science, with a minimum of collision between interests. It is more able financially than any possible combination of private investors. And it is impersonal, has but one motive—the most superior possible service.

The Government could finance the installation of receiving equipment with a minimum of difficulty, and the whole expense could be met by the relatively simple process of taxation. But is that all?

The transient holders of public office adore power, and do not forego it without pain. What man ever willingly surrenders his seat among the mighty? If men were incorruptible, if ideals were never contaminated, if absolutism were really absolute and dependably moral, then the simple, efficient device of governmental production, distribution, and maintenance might serve for television and everything else. Do you think such a state of affairs is possible? And would you risk a civilization's future in a gamble for such perfection?

Appendix A

TELEVISION BROADCAST STATIONS 1937

LICENSEE AND LOCATION	CALL LETTERS	FREQUENCY KC OR GROUP	POWER VISUAL	AURAL
Columbia Broadcasting System, Inc., New York, N. Y.	W 2 X A X	B, C	50w	
Don Lee Broadcasting System, Los Angeles, Calif.	W 6 X A O	B, C	150w	150w
Farnsworth Television Incorporated of Pa., Springfield, Pa.	W 3 X P F	B, C	4kw	1kw (C. P. only)
First National Television, Incorporated, Kansas City, Mo.	W 9 X A L	B, C	300w	150w
General Television Corporation, Boston, Mass.	W 1 X G	B, C	500w	
The Journal Company, Milwaukee, Wisconsin	W 9 X D	B, C	500w	

LICENSEE AND LOCATION	CALL LETTERS	FREQUENCY KC OR GROUP	POWER VISUAL	POWER AURAL
Kansas State College of Agriculture and Applied Science, Manhattan, Kansas	W 9 X A K	A	125w	125w
National Broadcasting Co., Inc., New York, N. Y.	W 2 X B S	B, C	12kw	15kw
Philco Radio & Television Corp., Philadelphia, Pa.	W 3 X E	B, C	10kw	10kw
Purdue University, West Lafayette, Ind.	W 9 X G	A	1500w	
Radio Pictures, Inc. Long Island City, N. Y.	W 2 X D R	B, C	1kw	500w
RCA Manufacturing Co., Inc., Portable (Bldg. #8 of Camden Plant)	W 3 X A D	D(124,000 to 130,000)	500w	500w
RCA Manufacturing Co., Inc., Camden, N. J.	W 3 X E P	B, C	30kw	30kw
RCA Manufacturing Co., Inc., Portable-Mobile	W 10 X X	B, C	50w	
The Sparks-Withington Company, Jackson, Mich.	W 8 X A N	B, C	100w	100w

APPENDIX A

LICENSEE AND LOCATION	CALL LETTERS	FREQUENCY KC OR GROUP	POWER VISUAL AURAL
University of Iowa, Iowa City, Iowa	W 9 X K	A	100W
University of Iowa, Iowa City, Iowa	W 9 X U I	B, C	100W
Dr. George W. Young, Minneapolis, Minn.	W 9 X A T	B, C	500W

GROUP A	GROUP B
2000 to 2100 kc	42,000 to 56,000 kc

GROUP C	GROUP D
60,000 to 86,000 kc	Any 6000 kc frequency band above 110,000 kc excluding 400,000 to 401,000 kc.

The low definition group (2 to 2.1 megacycles) is made up wholly of noncommercial licensees seeking to develop service for rural areas. In general these use mechanical scanning systems of about sixty line, twenty frame definition. Programs have been received as far as three thousand miles from transmitters.

The high definition operators (42 megacycles and up) are concentrating on intensely populated areas to which they expect to offer programs of an elaborate nature. In general, they have service areas of less than fifty miles radius, and use both mechanical and electronic type scanners, of four hundred line, thirty frame average definition.

145

History!

BLOOMINGDALE'S PRESENTS TO AMERICA THE FIRST PUBLIC FASHION SHOW

by Television

Simultaneously visible in four places on our third floor—telecast from our sixth floor studio

As merchants we conceive it to be a part of our responsibility to the public to explore every good device for the presentation of our merchandise. After months of experimentation, television, the modern miracle, is sufficiently developed to indicate its future service to advertising and fashions. Today we present a fashion show of millinery on living models, staged in one part of our store and simultaneously visible through kinets in four places in the building. No mirrors, no rabbits, no black top hats. Television! We invite you to see the present accomplishment—experimental and explorative—of one of the great discoveries of our time.

TELEVISION FASHION SHOW will be given at 10 o'clock this morning and every half hour thereafter. It will be simultaneously visible in four places on our third floor and a limited number of people can watch the actual televising in our sixth floor studio. This is being done with the cooperation of the American Television Corporation.

THIS IS A VIDEO CAMERA

It is the camera that picks up and transmits the image—the instrument by means of which you will see through the kinets some of our most alluring millinery projected through space, tomorrow. The word "kinet" will soon become part of the parlance of the times.

SHOP UNTIL 9 TONIGHT!

BLOOMINGDALE'S · LEXINGTON AT 59TH STREET · VOLUNTEER 5-5900

TELEVISION IN STORE CARRIES HAT STYLES

New Merchandising Method Is Seen in Special Wire Set-Up

A television showing of a millinery fashion show yesterday at Bloomingdale Brothers, Inc., was characterized as "a peep into the future of merchandising" by I. A. Hirschmann, sales and advertising director, who prefaced the demonstration by outlining the possible uses of television in a department store.

As the manikins approached the electric camera in a studio on the sixth floor of the store wearing the latest styles in hats, their images were carried by wires, limiting reception to the building, to "kinets," or viewing screens, seven by nine inches square on the third floor, where the audience was gathered. A running commentary accompanied the showing describing the hats in detail.

The manikins were illuminated by a powerful floodlight and wore a sun-tan make-up, while their eyes and lips were painted a deep blue to facilitate reproduction. Only their heads and shoulders were visible to the audience.

It is planned to install the "kinets" at escalator landings and in various departments of the store to encourage multiple sales from customers who come with a fixed idea of what they want to buy. For example, while a woman is being fitted for shoes she can see the latest styles in dresses on the television screen and may be encouraged to buy one. The camera also has an attachment for film and therefore can transmit a continuous fashion show.

Beginning this morning at 10 o'clock, the millinery fashion show will be exhibited every half hour on four receivers on the third floor of the store.

HEAR the opening of the N. Y. WORLD'S FAIR VIA RADIO
(VIA MAJOR RADIO STATIONS)

SEE President Roosevelt open the N. Y. WORLD'S FAIR VIA TELEVISION
(VIA NBC TELEVISION STATION)

This Sunday, at 2:30 P. M. President Roosevelt will officially open the N. Y. World's Fair and, at the same time, a new industry TELEVISION will be launched . . . for in addition to being broadcast over regular radio channels, the opening ceremonies are being televised by NBC. Thus, after years of experimentation and research, Television comes into its own.

SEE THE RCA VICTOR OR DUMONT TELEVISION RECEIVERS AT THESE DAVEGA STORES

Downtown—65 Cortlandt St.   Times Square—152 West 42nd St.
Madison Sq. Garden—825 Eighth Ave.   Empire State Bldg.—18 West 34th St.
Newark—60 Park Place (Military Park Bldg.)

1939 RCA Victor
RADIO ADAPTED FOR USE WITH TELEVISION ATTACHMENT

Save $30.00

SWITCH KEY ADAPTED FOR USE WITH TELEVISION OR VICTROLA

Manufacturer's List Price 87.50

$57.50

Outstanding Features for Fine Performance

- American and Foreign Reception
- Improved Electric Tuning
- Continuously Variable Tone Control
- RCA Victor Metal Tubes
- Automatic Tone Compensation
- Victrola Push-Button Switch
- RCA Victor Magic Eye
- Console Grand Cabinet

**NO MONEY DOWN
NO INTEREST TO PAY**

on Davega's 90-Day "Charge-It" Plan. Or take as long as a year to pay on Davega's Budget Plan—Small carrying charge.

149

TELEVISION
Gives its "Coming-out Party" Sunday

ON THEIR DU MONT TELEVISION RECEIVERS

...many leading stores invite you to *SEE* and hear

the opening of the World's Fair next Sunday

(Telecast begins at 12:30 P.M., April 30th, 1939.)

NEW YORK looks forward to next Sunday as its proudest day; the day when Mr. Grover Whalen will ring up the curtain on the world's greatest World's Fair.

But April 30th, 1939, will glisten in the pages of history for an entirely different reason. It will be commemorated by the world of tomorrow as the day when Television gave its first real "coming-out party" in the U.S.A.

At 12:30 P. M. next Sunday, the National Broadcasting Company launches a new Telecast series; with a program to cover the World's Fair Opening; featuring a sight and sound address by the Nation's President.

Seated in easy chairs before DU MONT Television Receivers, and in many leading stores, scores of New Yorkers will see and hear a three and one-half hour panorama of these dedicatory ceremonies. They will eye-witness and listen to each and every exciting event as it occurs; seeing and hearing it as clearly as though they were on the grounds . . . milling with the crowd.

For this magic privilege of witnessing the Fair's dedication while comfortably seated before a Television Receiver . . thanks must go to DU MONT. DU MONT was first to place a practical Television Receiver on the market, making home sets available to New Yorkers as early as last October.

No one man or group of men can claim the glory of discovering Television. A long line of scientists have been exploring its possibilities since the early 1880's.

But no one man or group of men has done more to make Television the reality it is today than has Allen B. Du Mont . . . in his pioneering and development of the Cathode-Ray Tube . . . the very heart of Television reception.

Television is here.

SOONER THAN YOU REALIZE it will play a vital part in the life of the average American, enriching his daily opportunity to SEE and hear what's going on in the world.

SOONER THAN YOU REALIZE . . . regular Television schedules will include "on-the-spot" Telecasts of every major event in the News, in Sports or in the Entertainment field.

Whether you are interested in Television for today or tomorrow . . keep this in mind. DU MONT was FIRST in placing dependable Television Receivers on the market. And DU MONT intends to remain FIRST in bringing the highest quality Television reception into the home.

SEE
DU MONT TELEVISION
AS WELL AS HEAR

First in   the Field

© Copyright 1939 by Allen B. Du Mont Laboratories, Inc.

FIRST IN THIS FIELD
WITH RECEIVING SETS
ON DISPLAY AT LEADING STORES

DU MONT TELEVISION RECEIVERS

NEW YORK OFFICES
ALLEN B. DU MONT LABORATORIES, INC.
515 MADISON AVENUE

RADIO STRAPS ON ITS CAMERA AND GOES TO THE FAIR

TODAY'S EYE-OPENER

Telecast of President at the World's Fair To Start Wheels of New Industry

By ORRIN E. DUNLAP Jr.

The mobile tele-van that will roam the Flushing flats to tele-picture the New York World's Fair.

Top: Those at home who see the opening of the Fair today by television will look through a "window" on the air, behind which twenty-four Nipkow discs transform the invisible into sound and scenes.

The aerial in the garden of RCA's radio tube-shaped building will relay scenes of the Fair to the main station atop the skyscraper on Manhattan Island.

Top: The "feather" in the steel cap of the Empire State Building is the latest television aerial which sprays sound from the circular affair at the top of the mast and pictures from the ballet-shaped fins below.

NEWS FROM THE MARKET

Television Now Drops Mantle of Mystery And Public Becomes Its Judge

ADVICE TO PARENTS ON 'CHILD RADIO'

WPA Gives Some Hints For Family Harmony In the Home

FAIR ON THE AIR

BEHIND THE SCENES

Radio Opera Reveals the Need of Melody To Enchant the Unseen Audience

By G. C. E. D. Jr.

Tonight!

KIX AND WARNER BROS. PRESENT

"THE GROUCH CLUB"

Newest Comedy Show on the Air

★ Starring Grouch Master, Jack Lescoulie; Announcer Nelson Case, author Brian Lane Leonard and his Warner Bros. Dramatic Group

WEAF · 6:30 P. M.

SPONSORED BY GENERAL MILLS

Telecast of President at the World's Fair To Start Wheels of New Industry

By ORRIN E. DUNLAP Jr.

WITH all the exuberance of a boy with a new kodak, the radio men pick up their electric cameras today and go to the World's Fair to televise the opening spectacle and to telecast President Roosevelt as a "first" in this new category of broadcasting.

For Long Island's north shore commuters, among them many radio men who have passed across the Flushing flats daily for years, the new wonder city seems like a dream. It is almost unbelievable that television as a new industry is to take off from the desolate site which only a few years ago looked like a hopeless mountain of ashes and debris, known as the meadows, the swamp and the dump.

Even Dr. Vladimir Zworykin, inventor of the iconoscopic eye, who for years has lived at Forest Hills, never dreamed five years ago that he would see television start as an industry almost in his own backyard. It's the old story, "acres of diamonds," coming true again.

While officials of the Fair are planning to handle 1,000,000 visitors today, radio microphones will tell the story of the scene to a greater audience stretched across the country and in foreign lands, making it a World's Fair in the air as well as on Long Island. And while the throngs mill around the midways and try to catch a glimpse of the President and other dignitaries amid the pageantry of the exposition, those few televiewers who stay at home in the metropolitan area may see even clearer by television.

The cameras with the telephoto lenses will poke closer than visitors can hope to push, and, therefore, television, with its reputation for intimacy, will reveal what advantages, if any, are promised for those nongregarious souls who sit comfortably at home looking through space with a length of antenna wire as a new sort of telescope. The weather won't matter for the televiewers.

* * *

IN the chronology of radio "April 30, 1939," becomes a historic date in the annals of science. It holds the same unforgettable significance as "Nov. 2, 1920," generally considered as the date when broadcasting started as an industry in the smoky air around Pittsburgh.

Television has been coming 'round the corner for many moons. Now and then it has peeked around in various demonstrations to look for its cue to step out on the world as a stage. Today the curtain goes up, not as some thought it might, on the glamor of a setting in Hollywood or on an elaborate scenic studio in that electrical acropolis, Radio City, but on what was once an ash dump on Long Island. But on the ashes from the stoves and furnaces of many homes has arisen "the city of tomorrow." Television is part of it.

That thousands and thousands will tread the pathways of the exposition today and in the days to come is evidenced by the fact that the advance ticket sale is 3,000,000.

There is no such yardstick for television. Its audience, like that of radio, is elusive, hidden and uncountable with any accuracy. There may be 200 tele-radios in operation within a fifty-mile radius of Manhattan's spires, but there is no gauge or turnstile in the air, as at the Fair, to clock the number of sightseers. Possibly 500 will look in on the opening when the curtain rolls up, but that is only a guess.

* * *

FOR the television showman the Fair offers a challenge of perpetual motion. If he is to follow cinema theatre practice his life will not be so hectic, because the playhouses are open for a limited number of shows daily, but if he tries to follow in the steps of broadcasting he is likely to die young.

The cameras like the microphone have an insatiable electrical appetite and if operated every minute from morn to beyond midnight they "eat" up much in sound and pictures. The ear, of course, as eighteen years of broadcasting has proven, will tolerate repetition in music, but the more fickle and discerning eye will not look at the same picture or performance over and over. It is too easy to drop the eye lid or turn the head; the ear has no such guards.

So for the television program planner this day is a day of challenge. He has worked and planned for weeks for the telecasting of the Fair, and by night when it is all over, he may suddenly realize that tomorrow is fast approaching and so are days and days on end for which he must grind out pictures if television is to survive as an industry. The showman is the man who must fan the spark the research experts have lighted; whether he can blow it into a flame that will sweep the country as a "craze" as did broadcasting in the Coolidge era, remains to be seen.

If the cameras click as they should today, President Roosevelt will long be remembered as the first Chief Executive to be seen on the air, as Harding is on the record as the first to take up the microphone while in the White House.

The very second the Presidential party motors into range of the "eyes" he will be on the air. The microphones have always waited until the stage was all set, until the exact second arrived for the show to begin. Television promises to be less formal, far more intimate because it will pick up sidelights and intimacies such as side chats. On the radio the President may cough and clear his throat before beginning the broadcast, but television is not likely to hide such preliminaries; it will be more natural, and if the President blows his nose or takes a drink of water during the telecast that will add to the naturalness of it all.

For weeks the mobile vans, television stations on wheels, have been practicing at the Fair grounds. They have been tossing pictures from Flushing to Radio City, and even in the rain and through low-hanging Long Island fogs, they have flashed remarkably clear scenes over the eight-mile bee-line. Hopes run high because the preview tests have been successful.

* * *

THE curtain goes up at 12:30 P. M. on the first act described as "the beginning of American television broadcasting." The program, on the air for three and a half hours, will include the parade and President Roosevelt delivering his address at the Federal Government Building, formally opening the Fair.

Then the ethereal projectors will switch to Radio City for a telecast made up of films. Thus television goes down the runways of space, as production belts of factories begin to move a trifle faster this week in sending new streamlined tele-radios to the home.

Regular evening programs to be telecast on Wednesday and Friday of each week, begin this week. By no means do the cameras plan to follow the pace set by the "mike"; they will be on the air from the

Radio City studio for sixty minutes beginning at 8 o'clock.

In addition to revealing how the cameras can work outdoors, the schedule calls for telecasting with Old Sol supplying the illumination on Wednesday, Thursday and Friday afternoon. Scenes will be relayed by the mobile units to the main transmitter atop the Empire State Building. The station's call is W2XBS; the picture travels on 45.25 megacycles; sound on 49.75.

* * *

THE transmission schedule also calls for twenty-three hours of film a week for the benefit of exhibitors at the Fair and dealers demonstrating home-sets. This timetable is Monday, Tuesday and Thursday from 11 A. M. to 4 P. M., and on Wednesdays and Fridays from 4 to 8 P. M., after which the regular evening performances will be picked up in the studios. The majority of programs will comprise ten-minute transmissions at fifteen-minute intervals, and the showmen warn "that the subject matter will be found repetitious, and the schedule may not be strictly maintained."

And so by sunset tonight television will have come from around the corner in quest of its destiny; to find its role in the art of amusing Americans, and to fit in with the social life of the land. Englishmen, who have been roaming outdoors with tele-cameras since the coronation of George VI are known to be watching what American enterprise and ingenuity will do with television once it gets "the bit in its teeth." They predict progress, but it is likely to be a slow, costly process, for television is a complex scientific giant that will take some time to get going full steam, especially with the instruments priced from $200 to $1,000. In mass production is television's hope for growth, lower prices and an audience.

14,000 TELE-SETS

FOURTEEN thousand television sets are in use in the London area, according to Radio and Electrical Marketing.

The population of London is 8,201,818. Telecasts have been on the air there for three years. The potential audience is estimated at 28,000 to 30,000.

154

Television Now Drops Mantle of Mystery And Public Becomes Its Judge

INQUIRY among stores planning to handle televison receivers reveals that there has been no rush on the part of New Yorkers during the past week to have video sets installed in time to look in on the opening ceremonies of the World's Fair today.

While the retailers have noted an increasing curlosity in television receivers among store patrons, they believe that there has been no rush of orders for several reasons: regular telecasting schedule to the home will not begin until Wednesday, sets will not go on sale officially until this week, advertising campaigns have not gone far, prices are high and the public is adopting a "watchful waiting" policy to see what television has to offer.

Estimating the Sales

Another factor of sales resistance, according to retailers, is wonderment on the part of the public whether or not present sets will soon be obsolete, and whether this will mean price-cutting later in the year on the models now being introduced. Radio manufacturers report, however, that production schedules will be geared to demand and the overproduction ills that upset the radio industry in the beginning will not be repeated in television.

It is estimated that today in the New York area there may be from 100 to 200 tele-sets, the majority in the hands of experimenters, engineers and officials of the broadcasting companies. How many televi-sion sets will be sold in 1939 is an open question in radio circles. Guesses range from 10,000 to 100,-000, with high hopes placed on "a television Christmas" if general business holds to keel. The less optimistic point to the fact that in London it has required three years of telecasting to get 14,000 sets "on the air."

The majority of the leading manufacturers have yet to introduce their tele-models, and in trade circles it is believed that the television campaign will not get under way with full steam until after Labor Day. Throughout the Summer the public interest will be studied. Furthermore, by that time the Columbia Broadcasting System's station atop the Chrysler Tower is expected to be on the air, adding to the program service.

The bulk of the country, however, is televisionless, so that it does not mean much to radio dealers outside of the New York region.

To pave the way, progressive stores are building air-conditioned sound proof demonstration booths that can be darkened.

Commenting on "how television will be sold," Radio and Television Today reports: "Because of the great problems of installation in apartment houses and the time and cost involved in locating the best spot for the antenna, so as to deliver the signal to the televisor strong and free of interference, almost every dealer turns thumbs down on home demonstrations.

"Most store owners, sales managers and company executives are of the opinion that in the early months of selling television there will be no price cutting to force sales volume, but many dealers express the view that there may be plenty of cut prices later, and worse headaches, resulting from changes in transmission methods or picture size, which would obsolete not only those televisors sold, but those in stock as well."

Fate Rests With Public

Marking "April 30, 1939" as the day on which television drops the mantle of mystery, Radio and Television Retailing, comments:

"From now on, television will be in the open, where the public can see how it works, can see what it will and will not do, can appraise its capacity, its state of perfection and its virtues. From now on, television will have to take its place with radio, phonographs, movies, as media of entertainment, and Mr. and Mrs. Consumer are going to be the final judges of its actual worth.

"Thus the question of what the arrival of television means to the radio dealer at this writing must still remain a matter of conjecture, because first the public must have a chance to react."

Probable Influences of Television on Society

By David Sarnoff

President, Radio Corporation of America, New York, New York

I.

To DISCUSS the effect which the general introduction of a new scientific development may have upon society is like attempting to predict, at his college commencement, the future career of a young man whom one has known from childhood. In the radio industry we have known television in this way, in a sense, for many years. We have watched its progress through the laboratory, and have encouraged it with financial and moral support while it struggled with the creation of sound fundamental principles upon which the future matured art must be based. We observed the coming and going of numerous irrational fancies, in methods and devices, during its adolescence.

Finally television is becoming adult, integrated in its technical standards and practices, with a cathode-ray system which has been uniformly adopted by the engineers of the radio industry. Yet, its maturity thus far has been attained only within the protecting confines of the homes and workshops of its scientific guardians. This Spring it graduated from this protective environment, and was thrust out to make its way in the world. What will it offer to mankind, and what response will mankind make to it?

Ever since the beginning of time man has sought to extend the power of his senses and to enlarge his capacity to perceive and respond to the world around him. Until a few centuries ago these instinctive strivings could utilize only the limited powers of the normal human senses and bodily capacities, unaided by scientific devices; so that adventurers wandered over the face of the earth on arduous journeys, and sailed the seas in lonely ships, to learn more of the nature of the world and to come into contact with the inhabitants of far-off places.

Writing and printing first enabled men to extend the power of their voices in order to communicate thoughts. With this extension of the powers of speech and hearing, the less gifted and less adventurous spirits could begin to apprehend the experiences and philosophy of those who were more fortunate in their mental or physical capacities, and thus satisfy vicariously their own instinctive desires for wider participation in the world's affairs.

Then, only a few hundred years ago—a tiny fraction of time compared to the ages that went before—man stumbled upon the scientific method. Inventive minds began to use basic scientific laws and principles for constructing devices which would fulfill many ancient human yearnings. Machines were made which multiplied many-fold the capabilities of human hands and muscles; optical instruments began to enhance the power of vision; railroads, steamships and automobiles increased the powers of locomotion and gave people the means of satisfying more readily the age-old desire for visiting other lands and places; communication devices brought distant friends and relatives close together.

One by one the shackles that chained man to the limited sphere of his own mind and his immediate neighborhood have been struck from him. Today he can move his body about rapidly, easily and at will; he can enlarge the powers of his hands and arms thousands of times; he can extend his voice by radio to other men throughout the world, and hear them in return. Now the last shackle is about to be broken; through television his eyesight promises to become all-embracing and world-wide. And not only is he given the power to see at great distances those things which may be evident within the limited spectrum of the visible rays of light, but also those which heretofore have been invisible because they could only be perceived through the use of waves outside the visible region.

II.

With the advent of television a new force is being given to the world. Who can tell what the power to extend vision will mean ultimately in the stream of human life? Could anyone have foreseen the vast social effects of electricity inherent in the voltaic cells of the early physicists or in the experiments of Faraday? Could we have foreseen that technological unemployment would be seriously regarded by some as a social consequence of the evolution of tools from the primitive axes and knives of our ancestors to the complex labor-saving devices of the present day? The most audacious imagination could not have envisioned the vastly ramified applications of electronic devices which have grown out of Edison's first observation of electron emission from the heated filament of a lamp.

Thus, it would require courage indeed to attempt to estimate the ultimate effects of television and all the scientific or social consequences that may flow from its introduction. We know only that inventions which gave us new powers have had far-reaching results in the history of the human race. Professor W. F. Ogburn has made a special study of the social effects of inventions, and in one of his papers he has pointed out some striking instances. For example, it is said that the use of gunpowder was a powerful factor in breaking down the system of life built around the feudal lord and his castle; the use of steam in connection with machinery greatly changed family life by taking industrial production out of the home and into the factory; important inventions of the past fifty years such as the telephone, the automobile, the airplane, the motion picture and radio are producing far-reaching effects on the family, government, education, industrial production, the habits and beliefs of people, and the economic well-being of nations. The social effects of inventions such as the airplane, radio and rayon have only just begun, comparatively speaking, and the effects of the telephone, the automobile and the motion picture are far from being completed.

Furthermore, Ogburn shows that an invention has primary and derivative effects. He says:

"The primary effect of the 'power inventions' namely steam, gasoline engines and electric motors has been upon the economic or industrial organization of the family; women went to work outside the home, children were employed in factories, and the father ceased to be much of an employer or manager of household labor. There followed a shift of authority from father and home to industry and state. In cities homes became limited as to space, and more time was spent outside by the members of the family. In a similar way, inventions impinged upon government, because of the growth of large corporations for manufacturing and for providing services which were made possible through power inventions. The regulatory functions of government increased, and taxation methods were modified. Many more government activities were assumed or engaged in through the force of the circumstances created by the changed economic organization. Finally, another derivative effect occurred in connection with modifications of social views and philosophies. Attitudes toward a philosophy of laissez faire are undergoing changes as more and more governmental services are demanded. Attitudes toward recreation and leisure time change, with city conditions and repetitive labor in factories."

We have some basis for predicting the probable *primary* social effects of television by regarding it as an extension of the present system of aural broadcasting. Indeed, some of the sociologists who have given consideration to this subject believe that a new series of effects more important than those of sound broadcasting are not likely to occur. S. C. Gilfillan says in this connection:[*]

"In ordinary life the eye is used more in perception than the ear. It has been suggested, therefore, that visual broadcasting when perfected will have even more important effects than aural broadcasting. Such an idea must be accepted with caution. From the social standpoint, the most significant development took place when radio made it possible to send news, music and propaganda through the air into the home. Six years ago the authors of the chapter of 'Invention' in *Recent Social Trends* were able to list 150 social effects of radio in its aural form. It does not seem likely that television will introduce a new list of social effects which is longer or more important. Addition of sound to pictures doubt-

----
[*] S. C. Gilfillan, "Social Effects of Inventions," in *Technological Trends and National Policy*. Report of the National Resources Committee, Washington (1937).

less produced relatively few new social effects of the cinema. Addition of pictures to sound should be more important in the case of radio, because of the greater use of the sense of vision than of the sense of hearing. Yet it is likely that the main impact of television will be to intensify the social effects which broadcasting already is producing."

While on the whole, this seems to be a sound line of reasoning, nevertheless the degree of intensification of present-day social and psychological effects of radio may be such as to produce a new series of changes in human lives and habits. Television will finally bring to people in their homes, for the first time in history, a complete means of instantaneous participation in the sights and sounds of the entire outer world. It will be more realistic than a motion picture, because it will project the present instead of the past. Aural radio already has demonstrated the greatly heightened psychological significance, to the listener, of feeling that he is present at the radio performance, as a member of an audience listening to living performers. The sensation that one is participating in an event actually taking place at the precise moment of hearing it is quite different and much more intense than the sensation one has in witnessing a sound picture or hearing a record of the same event, later on. With the advent of television, the combined emotional results of both seeing and hearing an event or a performance at the instant of its occurrence become new forces to be reckoned with, and they will be much greater forces than those aroused by audition only. The emotional appeal of pictures to the mass of people is everywhere apparent. We have only to regard the success of motion pictures, tabloid newspapers and modern picture magazines, to be convinced of this.

III.

Let us consider next what sort of program material television may present to its audience. Radio programs today cover almost every conceivable type of material that may be of value as entertainment, instruction and news. But while the scope of television programs will be equally broad, it is clear that the relative emphasis on the various types of subject matter can be changed to advantage. In aural radio we tend to emphasize program material that may be enjoyed without the use of vision; hence music forms a major part of aural radio programs. In television it will be natural to emphasize types of program material where the addition of visibility will enhance the emotional effect—such as drama, news, or sporting events.

Radio already has made significant contributions to novel dramatic forms and materials. Experimentation is constantly going on, under the daily pressure of providing ever-changing programs. Famous dramatists, actors and producers are turning in increasing numbers to radio as a new and important medium, and the intellectual standard of much radio drama is in the best tradition of the legitimate theatre. With the advent of television, a new impetus will be given to this form of art, and we may expect it gradually to take the place of some other types of programs which now occupy a large part of radio time.

While some television dramas may be recorded on film, for convenience or for network distribution, it is not certain that the standards, methods or artistic ideas of the present-day motion picture industry will control the material presented. Radio has always been an independent force, and has broken new ground in what it has done. A first-class radio program is like no theatrical or motion picture presentation that ever was. It is a new thing in the world. Similarly, it is quite likely that television drama will be a new development, using the best of the theatre and motion pictures, and building a new art-form based upon these.

It is probable that television drama of high caliber and produced by first-rate artists, will materially raise the level of dramatic taste of the American nation, just as aural broadcasting has raised the general level of musical appreciation.

Advertisers who sponsor radio programs will be given new possibilities of appeal through the medium of television. We need not fear, however, that this will mean an increase in the amount of "sales talk." In fact, it is probable that pictures or demonstrations of the product with the briefest possible messages will take the place of the more extensive announcements which are necessitated by the limitations of aural radio.

Political addresses will be more effective when the candidate is both seen and heard, and is able to supplement his address with charts or pictures. Showmanship in presenting a political appeal will be more effective than mere skill in talking, or the possession of a good radio voice; while good appearance may become of increasing importance, with the audience observing the candidate in close-up views.

An outstanding contribution of television will be its ability to bring news and sporting events to the listener while they are occurring. The widespread public participation in events such as those which occurred during the European war crisis in the summer of 1938, and the intensity of the mass emotions aroused thereby, have given us a glimpse of the possibilities of this phase of radio. It may readily be imagined what will be the results when television adds to the effect of reality by projecting the vision as well as the hearing of the audience to the scene of action.

Some social scientists have pointed out the greater possibilities of propaganda when presented by television. The great mass of the human race is not critical, and temporarily, at least, may be swayed by appeals to the emotions rather than to reason. In European countries which have succumbed to dictatorships extraordinary changes have been brought about in a very short time, with the aid of radio propaganda, in the expressed beliefs and actions of vast populations. These have been led to accept whole ideologies contrary to their former beliefs, because of skillfully presented ideas which have been spread to every home in the land with the speed of light and with a minimum of effort. The advent of television makes it even more important than heretofore to preserve for radio broadcasting in our country the precious right to freedom of discussion, and to guard against its exploitation for transmitting propaganda intended to arouse destructive class struggles, racial animosities or religious hatreds.

Educational institutions are gradually adopting mechanical inventions as aids to teaching, and radio receivers as well as phonographs are becoming increasingly familiar sights in schoolrooms. Because of these, the children of today have heard immeasurably more good music, and are more keenly conscious of world history in the making, than those of the previous generation. The possibilities of sound motion pictures for vitalizing and dramatizing scientific subjects, geography and history have been demonstrated; but schools are slow to make use of these because of the expense of the films, and the lack of organization among the hundreds of thousands of school administrations where cooperation is necessary in such a large-scale undertaking. With television we may find the educational uses of radio increasing; for while children may be bored and restless when merely listening to a speaker without seeing him, living pictures will capture and hold their interest.

IV.

There is another aspect of television which is important, and this is the nature and effects of its by-products. New instrumentalities have been developed, specifically for the purpose of transmitting visual intelligence by radio. These include iconoscopes, or devices for converting a light image into electric currents, amplifiers of wide frequency range, high powered, ultra-short wave transmitters and kinescopes which reproduce the original image by converting electric currents into light. All these devices are beginning to find applications in fields remote from television, and, as familiarity with them grows, their fields of application no doubt will be extended.

The whole subject of electron optics, or the control of electron beams by electric and magnetic fields, has received great attention because of its importance in television devices. This has led to a whole new range of possibilities in optical devices. Applications of this to astronomy, and in other fields where weak or distant sources of radiation must be dealt with, are future possibilities.

Some of the fields in which these television devices may bring about important advances are in marine or aerial navigation, by permitting vision at night or in fogs through the use of infra-red rays; in metallurgical, chemical, physical and biological research; in manufacturing processes as substitutes for human vision or for control purposes; in national defense; for advertising or display use in department stores, in showing goods exhibited at a central point throughout the store or in show-windows; for personal or business communication in transmitting visual intelligence as we now transmit the voice by telephone; in printing and copying devices; in new photographic or motion picture devices where "light amplification" may be used to advantage; and in any other fields where an automatic, never-failing substitute for the human eye may be useful.

V.

I have suggested some of the more immediate possibilities in the effects upon society of the advent of television. What of the more distant future, or *derivative* effects?

It seems to be the general opinion of authorities on population trends that life in the United States several decades from now will differ in important respects from that of the present time. The chief events which are anticipated are a continued increase in leisure time, an increase in the average age of the population, and a greater geographic decentralization or distribution of industry. The application of television devices will affect and be affected by these occurrences.

The average length of the full-time week for industrial workers has decreased from nearly 60 hours in 1890 to less than 40 hours at the present time. Improvements in manufacturing methods have had the effect partly of raising wages and partly of decreasing the working week. There is every reason to expect a continuance of these processes, considered on a long-time basis. The combined ingenuity of the social and physical scientist, encouraged by a sympathetic government, should in time produce the much desired results of more pay, shorter hours of labor and longer hours of leisure.

At the same time, if the birth rate continues its present declining tendencies, the distribution of population in accordance with age will alter materially. Population experts have estimated that whereas in the 1930 census only 23 percent of the population was over 45 years of age, by 1980 we will have 38 percent of the population in this age group. The whole tendency will be towards a predominantly middle-aged and elderly population.

A decline in the population of large cities is expected by the National Resources Board to set in some time between 1945 and 1960, with people moving into "satellite" areas within the metropolitan districts. We have already observed how the introduction of the automobile spurred the development of suburbs of large cities. Now, with steadily cheaper cars, increased and improved highways, it is anticipated—and the tendency is already clearly evident—that rural communities within perhaps fifty miles of the cities will increase in population and develop in scope.

All this provides a picture of a population which may increasingly center its interests once more in the home; a population with ample leisure time, of predominantly mature years, and widespread distribution, in individual small houses which they will be able to afford because of the development of low-cost home construction and increased income per family. With such a setting, radio-television will be a vital element in the lives of these people. It may become their principal source of entertainment, education and news. It will link together in mind and spirit these vast numbers of individual homes, as the high speed automobile roads and airways will link them together physically.

We may also anticipate a rising standard of culture, with universal education of both adults and children. New York State is now considering the extension of the present high school courses to six years; if this plan is widely adopted, we will soon have the equivalent of junior college training established as the minimum standard for graduates of our public school system. In the distant future of which we speak, it may be assumed that most persons will have at least an education of this level. What this may mean in terms of the type of material to be broadcast, and its place in the cultural life of the community, is stimulating to the imagination.

We have seen how much the general level of musical taste in this country has already been raised by the widespread radio broadcasting of good music. People to whom such matters as grand opera and symphonic music were unknown fifteen years ago are becoming increasingly familiar with them. With television, a similar widening cultural development in appreciation of the best in drama, the dance, painting and sculpture may be expected. Through television, coupled with the universal increase in schooling, Americans may attain the highest general cultural level of any people in the history of the world.

What of the effects upon existing institutions, such as motion pictures, the theatre, schools and churches?

The motion picture industry may become an important source of supply of recorded programs to television broadcasters, where such recordings may serve the purposes of program material more conveniently than direct transmission of living actors. There are other possibilities too for cooperation between the motion picture industry and television. Each should be able to stimulate the other and this should result in an enlarged service to the public.

With a rising cultural level, we may expect also an increase in the number of creative artists working with the materials of the theatre. Such artists will be used not only by the television broadcasting systems; they will find additional outlets for their creative energies. Through these new developments we may see a rebirth of local community theatres for the production of legiti-mate drama, musical performances, dances, and the like.

The school systems will probably make increasing use of television as part of the educational program; for with this medium it will become possible for the best teachers in the land to give carefully prepared and illustrated lectures to millions of children simultaneously.

Church broadcasting will rise to new spiritual levels, for with television the listeners can participate most intimately in the services of the greatest cathedrals; they will not only hear the ministers and the music, but see the preacher face to face as he delivers his sermon, witness the responsiveness of the audience, and observe directly the solemn ceremonies at the altar.

Thus, the ultimate contribution of television will be its service towards unification of the life of the nation, and at the same time the greater development of the life of the individual. We who have labored in the creation of this promising new instrumentality are proud to launch it upon its way, and hope that through its proper use America will rise to new heights as a nation of free people and high ideals.

We Present
TELEVISION

Edited by

JOHN PORTERFIELD

and

KAY REYNOLDS

New York

W · W · NORTON & COMPANY · INC ·

Copyright, 1940, by
W. W. NORTON & COMPANY, INC.
70 Fifth Avenue, New York, N. Y.

PRINTED IN THE UNITED STATES OF AMERICA

Contents

Contents

List of Illustrations

List of Illustrations

List of Illustrations

Editors' Foreword

IT IS fitting that the men who have struggled to inaugurate television as a public service should tell the story of its experimental stages of development. Herein they have written an informal record of this new industry and their hopes and visions for its future. This collection of experiences cannot be called a complete history, as the story of television is being written, day by day, in the studios by the engineers and artists who are constantly further developing and perfecting this new medium of expression. At the present time and for many years to come the expansion and growth of this new art will be a matter of constant research and co-operation between engineers, program directors, and artists.

The television studio is a wilderness of apparatus, with blunt-looking and massive cameras and bewildering batteries of lights. Out of this chaos a new art has been born. A new technique must be built, as orderly in its workings and its principles as the mechanism upon which it is based. This is a tremendous task. To bring this complicated invention to the peak of its expression, to the maximum of its power, is a great project. And it

Editors' Foreword

is this undertaking that engages the minds and the energies of the men who wrote this book.

The editors owe a debt of thanks to all those who worked towards making this an accurate and complete report on the development of television in this country. We cannot mention here all those who by their advice and counsel were a constant inspiration during the preparation of this book, but we are particularly grateful to Mr. Leif Eid of the Press Department of the National Broadcasting Company, to Dr. W. R. G. Baker, Manager of the Radio and Television Department of the General Electric Company, and to Dr. Orestes H. Caldwell, Editor of *Radio and Television Today*. We realize that without their aid and the courtesy extended to us by the many others who contributed their efforts we would have been unable to obtain the complete picture of this new medium.

We Present
TELEVISION

Introduction

BY WALDEMAR KAEMPFFERT
Science Editor, New York Times

I

WHO invented television?

There is nothing more pathetic in the annals of technology than the patriotic attempt to claim for a particular country or a particular man the credit for having invented a machine that has changed life and industry. Historians of engineering and social scientists know full well that there are no "first" inventions, that every contrivance is an organization of principles discovered singly by scores of now dead scientists and technicians, that every explorer of the unknown has a technical heritage which he must exploit before he can advance.

The scope of the technical heritage of those who gave us television is set forth by Mr. Donald Fink in the chapter on the "Technique of Television" that he has written for this book. Even his necessarily cursory review of past efforts makes it plain that television is a composite of a thousand inventions and discoveries, most of which seemingly had no relation at one time to the electrical transmission and reception of what we see but which were ultimately related to one another. Since the world thinks more highly of what the patent lawyers call "reduction to practice" than of observations and experiments from which new physical theories are derived, far too much has been made of the crude success achieved by John Baird in transmitting something more than mere silhouettes in 1925. This is not to deny the importance of seeing relationships where none has seen them before. The point is that we do not have to wait for a unique genius to piece together the scores of contrivances that culminate in a typesetting machine, an automatic machine-tool, a printing-press. In his *Social Change* Professor William F. Ogburn lists 148 major discoveries and inventions which were made simultaneously and independently by two or more scientists or technicians. Had the list been extended to include innovations of secondary importance it would have taken a volume as large as an unabridged dictionary to contain it. Every patent lawyer knows that his clients must reckon with anticipations of their conceptions.

Invention is inevitable in the right social environment. "I have not made events," Lincoln is said to have remarked. "Events have made me." In this larger aspect it is not the genius who leads us to such new goals as television, but society that pushes him on. Our television system is but the fulfillment of a wish cherished for centuries. What were the crystal gazers of old tales but televisionaries of a sort? Unable to conceive of any mechanism that would disembody the sense of sight, they turned vainly to magic. It was always thus. Soon after the telegraph was invented the fulfilling of the wish to see through vast distances began. At first the inventors did not dream of seeing over a wire, yet they took the first step by trying to send handwriting and simple drawings by telegraph. A pen-and-ink picture of a man is no real substitute for the man himself. Still it is better than a verbal description in dots and dashes.

Not one of the pioneers in facsimile telegraphy was aware that he was laying the groundwork for a new branch of engineering which was to concern itself with seeing at distances that the eye could never span. The motion picture was another step forward. Here were life, action, the illusion of reality. But it was not until the phenomenon of photoelectricity was discovered that television ceased to be a Jules Verne fantasy and became at least a theoretical possibility. When at last the modern photoelectric cell was invented and with it a way of converting light into electricity and electricity back into light the first groping experiments could be made. Even then it was necessary to develop the intermediate art of

Introduction

phototelegraphy—the art of sending photographs over a wire and through space with all their half-tones and fine detail. Line by line a photograph at the transmitter was scanned; line by line the receiver reproduced the photograph on sensitized paper. Substitute a face for the photograph to be transmitted, speed up the process of scanning until it is impossible for the eye to detect the process, and television is achieved. The difficult evolution of this conception and its expression in antennae, coils of wire, photoelectric cells and vacuum tubes has been so fully set forth elsewhere in this volume that it is unnecessary to elaborate it here.

What strikes the student of social sciences and the historian of technology is the manner in which this dream of television became a fact. The early facsimile telegraphers, cinematographers, phototelegraphers and television empiricists all belonged to what may be called the laissez-faire school of invention. It was a school that educated itself by cut-and-try experimenting, that converted attics and cellars into laboratories, that had to hawk its patents until at last some imaginative capitalist bought them. Just as the laissez-faire school of economics gave way to planning so the laissez-faire school of invention is giving ground to planned research. And for the simple reason that the attainment of television demanded a deeper knowledge of theoretical physics than any garret inventor could possess.

On the record it appears as if Europe had outdone the United States in the development of television. Both Great Britain and Germany had made television a commercial reality some years before America. To be sure Americans had their experimental transmitting stations, and several hundred engineers were testing receiving sets for years. But it was not until the New York World's Fair of 1939 was opened that programs were regularly and systematically broadcast. There was no technical reason why the United States should not have had television contemporaneously with Great Britain and Germany. With manufacturing companies here and abroad exchanging scientific information there never was any fundamental difference in operative principle between television in the United States and Europe. Great Britain anticipated us because the British Broadcasting Company, a government corporation, decides what shall be broadcast to the ear or to the eye. Operating revenues and profits are paid out of taxes on receivers. Advertising is negligible.

If television has apparently lagged in the United States it is for economic and not for technical reasons. Wherever the American promoter turned he faced doubts—doubts now happily resolved about the financial feasibility of connecting transmitting sets on a national scale, doubts about making images talk and sing all the time or only part of the time, doubts about finding enough good program material, doubts about what the public wanted. It was assumed from the outset that television must follow the general evolution of sound broadcasting. What held American television back was the traditional method of making an invention pay dividends. Assuming that the method was the best, which is debatable, it was necessary to conduct research not only in the laboratory to develop a satisfactory system of transmission and reception but among the people in order to discover what types of programs might be launched with the best prospect of advertising support.

In Europe and in the United States practical television is a product of organized science. The apparatus of the independent pioneers bears no more relation to what we have than a ferryboat bears to the Queen Mary. Yet television stands apart even in the latter-day abandonment of the laissez-faire process of leaving the business of contriving to independent, individualistic experimenters. It is probably the first very great invention deliberately developed by organized research and born full-grown in most of its essentials. Only the incandescent lamp furnishes a parallel. To introduce it Edison had to build the Pearl Street powerhouse, connect that powerhouse with offices and homes, design generating machinery, junction-boxes, meters, fuses, switches—all the complex apparatus of a new industry. Though this was the astounding triumph of one man let it not be forgotten that there were steam-engines, dynamos and other electrical essentials available and that Edison, with twenty or more assistants, had actually organized research much as it is organized in an industrial laboratory today. The physicists and engineers who gave us television did not find accumulated essentials at hand. Transmitters, receivers, nearly every element from the antenna at the studio to the type of screen that would best reproduce images in the home had to be created. Not a detail was overlooked. Organized industrial research saw to it that television was a commercially practical means of mass appeal before a single receiving set was offered for sale.

There can be no doubt that television would have evolved even under the old laissez-faire system. But at what pace? Between Nipkow, who invented an early method of mechanical scanning, and John Baird, the Scotsman who demonstrated television by its means in a crude way in 1925, is a gap of forty years. Between the application of organized science to television and the construction of practical transmitters and receivers is a gap of only a dozen years. Time was telescoped. The television set of today is not perfect, yet it is perfection itself when we compare it with the broadcasting receivers of the nineteen twenties.

If we have television today it is because the task of bringing it to an acceptable stage was assigned to trained engineers and physicists, who worked in groups on specific problems. There were literally hundreds of these men, and they worked for twelve years on problems that ranged over the wide field of physics and chemistry. Among these hundreds in one organization alone thirty-five won the 1940 National Pioneer Award of the National Association of Manufacturers for distinction in invention and industrial development. Their work is described in 229 published papers, 671 unpublished engineering reports, bulletins and memoranda. All the accumulated knowledge of electricity, heat, optics, acoustics, atomic physics and physical chemistry was drawn upon. A mere recital of the demands made on the industrial laboratories would be enough to convince anyone that television as we know it even now would not have been so rapidly achieved had its evolution been left to the discontinuous and casual divinations and methods of whittlers and empiricists. In these inventive years a leading American company spent $9,250,000 on research and exchanged its technical knowledge with European organizations that were likewise attacking the problems presented by television. How much money has thus far been spent in America and Europe in enabling ultimately millions to watch plays and news events at home no one knows. The total probably exceeds $20,000,000.

II

The bringing of television to the commercial stage before ever a receiver was placed on the market is but half the story. The transmission and reception of images through space involves not only generators, antennae, vacuum tubes, coils, condensers, and screens, but programs, actors, stage techniques, sets, lights. What has been accomplished in "studio research" several chapters in this book disclose with a vividness that fascinates and that gives one the impression of beholding a new art in the process of flowering. Already television is fully equipped to present news events, games, athletic contests, drama, in a word anything that can be presented to the eye. The motion picture passed through no such period of private incubation. It was launched with all its imperfections and left to discover its own way of becoming effective.

As it stands today the studio technique of television is a hybrid. It has borrowed heavily from radio, the stage, the motion-picture play. Probably it owes more to the motion picture than to any other medium of expression for the obvious reason that television and the motion picture have much in common. Yet it is already evident that what we have today is but a primitive beginning. Severe restrictions are imposed on both script writers and actors because large movements are impossible on a screen that can easily be covered with a pocket handkerchief. Despite British experiments with grand opera and ours with Gilbert and Sullivan we must await the screen of wall dimensions (which means a colossal tube, or optical projection and enlargement, if electronic scanning is to be retained) for the proper presentation of *Tristan und Isolde* and *Aïda*. When television figures of almost life size fill the eye drama and opera will be unfolded with an epic sweep that is now unattainable.

The film play has demonstrated the readiness with which our senses adjust themselves to new theatrical proportions. On the screen of the motion-picture theater we behold men and women of gigantic size, singing and talking with voices of incredible volume. We accept such incongruity as readily as we accept it in the puppet play. Even now the voices that well out of the loudspeaker are out of all proportion to the diminutive figures that gesticulate on the small television screen. Yet we do not protest. "Nobody can successfully tear a passion to tatters on a stage screen seven and a half by ten inches," one critic has remarked. But the television screen in England as well as in the United States proves the contrary. It is the quality of the passion that is important, not the size of the screen.

All this being so one cannot help but wonder how many of the possibilities of television have been unearthed by the patient men and women who have studied public taste in the matter of programs. The plain truth is that the conclusions thus far drawn are too obviously based on radio and motion-picture experience. We may agree that if a King is crowned in London or heavyweights meet to determine which of the two is to be champion of the world in his class the television camera will be on the spot. And we may also agree that we shall profit by lectures on deportment, interior decoration, art, hygiene and a thousand other matters in which instruction can be given by sight and sound. But the plays thus far written for the television screen must be regarded as artistic inventions that have hardly approached the perfection attained by the engineers.

The production technique that has already been devised differs considerably from that of the film. A Hollywood director can take a scene over and over again until at last his exactions are met. The montage editor can snip out scenes that seem to him superfluous and resort to various accelerating devices if the action is too slow. Indeed the production of motion pictures has an elasticity that makes the attainment of technical perfection possible. But the television play? When the eye of the video camera is turned upon the stage and waves ripple out that will become images on a screen sixty miles away the players must be letter perfect, which means that they must have rehearsed from seven to fourteen hours or even more. Success or failure depends on one cast of the dramatic die. There are no "retakes."

The studies of public taste which have accompanied the technical development of apparatus have been made primarily to discover the kind of appeal that is likely to be supported by the advertiser. The history of poetry, the drama and the novel proves plainly enough that the public must see a new art form before it hisses or applauds. Prediction can never supplant performance. The public may even remain indifferent, as it did for years after talking pictures were introduced. The sheer novelty of television will also deceive during the formative period. When the first films were exhibited it was enough to present an express train rushing across the landscape to arouse interest and wonder; when the radio stations began to broadcast to the nation we were delighted to hear time signals from Washington and bedtime stories from remote cities. No analysis of the public taste indicates what the possibilities of television are and how television plays should be written. A new school of creative artists must spring up, a school which will familiarize itself with technical procedures of the studio and which will exploit them in unfolding the story of human strength and weakness, devotion and treachery. A new and powerful medium of mass appeal has been developed. Only in commercial use will its potentialities be discovered.

An artist usually conceives his creation in terms of a given medium. A painting or a photograph of the Venus de Milo is a poor thing compared with the actual statue. Shakespeare translated to the screen is unsatisfactory without the majesty of line that we associate with him, for which reason *Hamlet* was never successfully produced in the days of the silent film. Many a Broadway success is admirably interpreted on a film only because the dramatist had Hollywood in mind when he conceived his situation and wrote his parts. Since a true work of art will not endure translation into any other medium than that which suits it best Mr. Walt Disney's animated drawings of Mickey Mouse and Donald Duck are the most important contributions ever made to the screen. If these dicta are valid, television as an art medium must evolve in its own way even though radio, motion-picture and stage techniques are partly applicable to its needs.

The relation of the motion picture to the television screen is not yet clear. That films will be used for landscape backgrounds or to transport the beholder to some distant scene is evident already from the televised version of *The Farmer Takes a Wife*. It is the status of the motion-picture theater that is in doubt. Not so long ago the British paid a guinea a ticket to see the Boon-Danaher fight televised in three theaters which were packed to the doors. Is this evidence merely of a lack of home television sets or of a preference for large screens and consequently for more life-like presentations? Or is it evidence of the value of the suspense that always accompanies participation in an event and of the immense superiority of the living, televised present to a newsreel which, though only a few hours old, must by sheer necessity present the historic past? Moreover if news events can be televised in theaters so can whole plays be televised from Hollywood or London studios. Whether the British and the Germans are right or not in wiring theaters for television the conclusion seems inescapable that motion pictures and television are destined to be interconnected in some way. It is significant that years ago the largest company secured an important position for itself in the film industry and that at least one important film-producing corporation has acquired an interest in a television organization.

Introduction

III

A radio broadcasting station may be in action as many as 5,000 program-hours a year. It follows that news events, vaudeville sketches, lectures on art, science, travel, home economics and the like will take up but a fraction of television time. British experience has shown that the full-length play, meaning something that lasts at least an hour and a half, is more in demand than any "short." All this means that ultimately a ravenous appetite for scripts, actors, set makers, mechanics and other auxiliaries must be satisfied. The expense to be faced is almost terrifying. Translated into terms of running time on the screen a motion-picture play may cost from $1,000 to $35,000 a minute, with $1,000 representing about the worst that the public will tolerate. If we are to have every day a new television comedy and tragedy lasting an hour and a half, the studios incur an outlay that dwarfs anything with which producers are familiar. Whole acres must be given over to "lots" on which half a dozen companies rehearse coming productions. An army of artisans must be kept busy preparing sets. Even if we assume an average expenditure of only $10,000 a minute, viewing time, a studio must reckon with an expenditure of somewhat less than $1,000,000 if each day only a single fresh play an hour and a half long is to be presented with the opulence to which Hollywood has accustomed us.

Television promoters talk hopefully of keeping the cost of production down to $500 a minute or even much less. Even the well-informed authors of the chapters in this volume that deal with costs cherish the illusion that Hollywood will be outdone in thrift—Hollywood which has had a generation of experience behind it and which could not escape the spending of several millions on *Gone with the Wind*. There is no reason to suppose that, as the viewing-screen grows in size, as it must, television can dispense with expensive sets or with outstanding players who will demand and receive salaries that will eclipse even those that Hollywood is willing to pay. Moreover, there are the extraordinary changes that have marked the development of sound broadcasting. Fifteen years ago great artists played and sang for nothing before the microphone. Now they triple and quadruple their normal concert fees because a whole nation cocks an electric ear when they perform.

Some misgivings have been expressed as to television's ability to find enough writers, set makers and players to satisfy the hunger for screen entertainment. There seems to be no good reason for dismay. Rather should television be regarded as an extraordinary opportunity, a new outlet for the artistic urge. Oliver Wendell Holmes's remark that every one of us has at least one good novel in him may prove to be more than the breakfast table pleasantry of his Autocrat. Despite the literary avalanche of books, periodicals and newspapers that descends on us there is an enormous amount of dormant artistic and dramatic ability. It is safe to prophesy that not only will there be an avid demand for good writers and players but that creators of television entertainment will spring up spontaneously to fill it.

IV

The future of television lies largely with the advertiser. He made broadcasting what it is today. He will play a large part in making television what it is destined to become, unless the government intervenes and controls programs as well as the use of the ether. It is a question whether his "sponsorship" has been a blessing or a curse to broadcasting; for American radio offers the worst and the best—the worst when advertising banalities pour out of the loudspeaker, the best when Arturo Toscanini or Bruno Walter conducts a great symphony orchestra maintained by the National Broadcasting Company.

To the advertiser television will prove to be a temptation hard to resist. He will suit the action to the word, the word to the action. "I put a little toothpowder into the palm of my hand," an agreeable voice will surely say as its owner appears to the eye. "I spread it on the brush so-o—." How much of this televisionaries will stand remains to be seen. There is statistical evidence enough that in hundreds of thousands of homes the radio loudspeaker talks, sings and plays from eight to ten hours a day. Not that anyone listens intently every minute but that something about music and speech supplies a background to which we may or may not pay attention. Ma goes about her business in the household and Pa reads the paper while the radio blares music from a brass band in Washington or croons sentimental ditties from a New York restaurant or dance floor. Television is more than background. Either Ma and Pa look at the screen— or they do not.

Moreover there is the experience of the motion-picture houses. They have experimented with advertising lantern slides between drama and news and found that the public resented the assault on the eye and the mind. When the film was made to talk no one had the temerity to place advertising on the sound track. Nor has the film itself ever been widely used as an advertising medium in the theater. Even today the preliminary flashes that announce attractions to be projected next week are none too welcome.

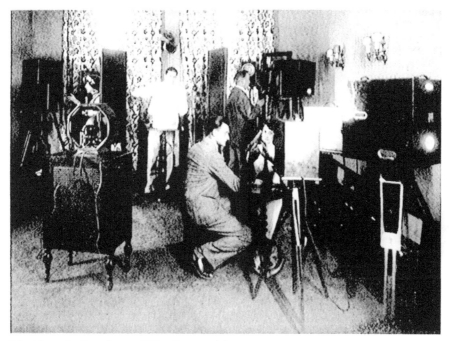

The first television drama, "The Queen's Messenger," telecast from WGY's studios, Schenectady, on September 11, 1928. (COURTESY GENERAL ELECTRIC CO.)

"Pirates of Penzance," first musical production telecast in NBC's regular television service on June 20, 1939, an example of the attempt of modern television to fuse the techniques of radio, stage, and motion pictures into a new art. (COURTESY NBC)

Illustration to show the difference between a fine and coarse halftone screen. (See page 70)

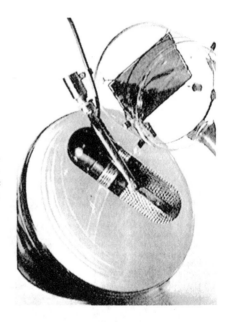

A symbolic arrangement of the iconoscope, the eye of television, the microphone, the ear of television, and the kinescope, television's reproducing tube.

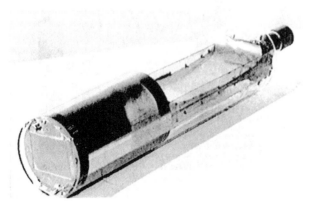

The orthicon, new television camera tube, which possesses many advantages over the older iconoscope. (COURTESY RCA)

Introduction

The telecasting of a news event or a play by some "sponsor" will surely be preceded, followed or interspersed with images of men and women smoking cigarettes (the package and trademark conspicuously revealed), smacking their lips over a drink of some new brand of coffee or ginger ale, extolling the virtues of a face cream, all to the accompaniment of advertising patter. Will the masses like it? They respond to printed advertising now. But they are volunteers. Television must change them into conscripts. Must we reconcile ourselves to forced visual advertising, just as we reconciled ourselves, though none too cheerfully, to the advertising banalities of the loudspeaker? Possibly we may be willing to pay the price of transitory boredom to see and hear a stirring play. It seems more likely that Ma will wash the dishes and Pa will read the newspaper until the announcer says: "The play is about to begin."

The chances are that in the beginning the advertiser will overdo television—that is drive viewers away by insisting on too long a display of his products and too much accompanying sales talk. Every minute on the television screen must be exciting. A new kind of advertising display, a new kind of dramatic action must be invented. Content is more important than advertisers think. Gilbert Seldes in an article which he wrote for the *Atlantic Monthly* of April, 1937, sagely remarks that as attention is intensified the shorter must be its duration. This means that television must keep us on edge, and that radio as we know it will seem slow in comparison. To drive home his point the advertiser must be dynamic and brief. He will have to cultivate restraint—to imply rather than to state directly what he wishes to convey. The most effective poster I ever saw was one displayed by the defunct North German Lloyd some thirty years ago. A Hindu fakir playing a pipe while a serpent rose and writhed, the words "North German Lloyd Oriental Cruises"—there was nothing more. But it was enough to send hundreds to the ticket office. Television advertising must be equally quick and provocative.

V

To predict a great invention's future is always hazardous. None of those who assisted in its early technical development seems to have realized that the motion picture was not merely an extension of photography but a means of entertainment and hence a new medium in which the dramatic artist could express himself. Edison banked heavily on the utilization of the phonograph for office dictation and only incidentally on the eagerness with which millions would listen to the records of great orchestras and singers. Indeed he even predicted that his invention would find one of its major uses in recording the last words of the dying. Morse and Field never foresaw the part that the telegraph and submarine cable would play in international arbitrage and in the creation of telegraph business. So far as the record indicates Marconi and other early experimenters with Hertzian waves did not realize that wireless telegraphy would make it possible for fishermen to inform themselves about the demand for cod or halibut in Liverpool or New York or for the masters of tramp steamers to learn from their home offices where profitable cargoes could be picked up. When the telephone was introduced bankers protested that it was impossible to talk about money over a wire. Now some of the largest telephone switchboards in the world are to be found in Wall Street. It never occurred to Bell that some day we would ring up San Francisco from New York for no other purpose than to wish a friend well on his birthday. To Edison and his contemporaries the electric incandescent lamp was simply a competitor of the naked gas flame. Flood-lighting, the use of safe lamps in new ways on the stage and in the home, the introduction of little bulbs in surgery and medicine—these potentialities were not foreseen. In short, it is not the inventors or even their financial supporters who find all possible uses for new inventions but the public. The most glaring example of all is furnished by broadcasting. The publicity of radio telephoning, the ease with which anyone equipped with a receiving set could eavesdrop, was considered an unavoidable defect. It became an asset when Harding ran for election. And it was the public and not the engineers that made the discovery.

Despite the technical difficulties encountered in sending television images by wire, who can doubt that some day we shall not only talk by telephone to a business acquaintance or a friend but that we shall also see him as we talk. "I called him up this morning," we say now. Soon it will be, "I looked him up at his ranch in California." George Bernard Shaw made the most of the possibilities in his *Back to Methuselah*. Documents will be displayed and read across the continent. There is even more than a possibility that a housewife who shops by telephone will ask a department store to hold up to the video camera the stewpan that she thinks of buying or the butcher to exhibit a chicken in which she is interested.

The time is undoubtedly coming when the twelve directors of a corporation in twelve widely separated parts of the country will hold a meeting in the office of the chairman of the board without leaving their desks. Their opinions and votes rather than their physical presence is required. But inasmuch as there are always opposing forces on every board the soft-spoken word may not be enough. Like poker players, directors insist on seeing one another. So, a few decades hence, a meeting may well be a meeting of electrically disembodied personalities. The chairman sits in the usual, very dignified, funereally upholstered room at the usual flat-top desk graced by the usual framed photograph of the usual wife and children. At the far end of the room are eleven television screens. In the office of each of the other directors are eleven identical screens. In twelve different offices twelve voices and images, twelve electrical ghosts confer. Each of the twelve sees eleven faces before him. The chairman talks to Stewart C. Dobbin as if Mr. Dobbin were physically present and listens to Mr. Dobbin's objection to reducing the dividend from 6 to 5 per cent. And he knows that Dobbin is watching him and following his argument in opposing the reduction just as closely.

The spectacle of the officials of the Federal Reserve Bank, the Reichsbank, the Banque de France and the Bank of England traveling hundreds of miles to meet in London or Paris will pass with other quaint and cumbrous customs of the early twentieth century. Even parliaments may be reduced to picturesque relics of a past when legislatures were compelled to meet in a given place. Filibustering will hold no terrors. Bored members will simply go to bed.

Some day even the images of the home television receiver will be as large as those that smile and dance on the screen of a motion-picture theater. They will appear in all the colors of nature. They will lack a third dimension—solidity. Enormous as the engineering difficulties of supplying color and stereoscopic fidelity now seem nothing is impossible. We shall see something that apparently has thickness, depth, density. That something will be as real as if we beheld the three-dimensional, solid original through a window at the transmitting station. Yet if we touch it our fingers encounter only a wraith.

It would be strange indeed if television did not have its offshoots. Out of the photoelectric cell came a thousand ways of supplanting human hands, eyes and even brains in the factory; out of the vacuum tube came not only the radio broadcasting receiver but a way of following sensory impulses along nerves and even the new branch of medicine called electroencephalography which has made it possible to amplify electric waves that come from the brain and thus for the first time in history to study cerebral physiology without opening the skull. And now television promises us in turn electron telescopes and microscopes.

Introduction

There can be no doubt that with the growth of electron optics the frontier of science will be immensely widened. Light has its limitations. It is too coarse to reveal many objects that now escape microscopic or telescopic detection. Despite the new theories it is still convenient to regard it as a movement of waves in a hypothetical ether. Just as a rowboat is lost to view in the trough of enormous ocean waves so many an infinitesimal detail is lost in much larger waves of light. Even before television became a commercial reality the physicists were quick to seize upon the work that had been done to bring it to perfection. Here were beams of electrons painting images on screens—beams focused and controlled by electromagnetic fields which were the counterparts of lenses. The whole mechanism of television has only to be redesigned and readapted to meet new requirements so that electrons, which are far smaller than waves of light, may reveal what still remains invisible.

Some twenty diseases, among them infantile paralysis, smallpox, the common cold, are caused by viruses. But what are viruses? No one knows. Perhaps living organisms beyond the ordinary microscope, though that now seems unlikely in view of the work that Dr. Stanley has done in crystallizing the viruses of some plant diseases. The electron microscope may supply the answer. And what are genes, the invisible entities that determine whether we shall be tall or short, black or white, blue-eyed, or brown-eyed? Again the electron microscope may answer. And so with the examination of metals and their alloys, of cells in animal and plant tissues. Chemical analysis is not enough. Structure is important. And it is structure that electron optics will illuminate, whether it be of a crater on the moon, a virus, a microbe.

VI

It took Julius Caesar twenty-six days to send a letter from Britain to his dear friend Cicero in Rome, and this by the fastest courier in the army. Since he was not only a soldier but a literary man, Caesar did his best to make his message live. Words were chosen to make Cicero see events as Caesar had seen them. And so it has been with all the poets, romancers, historians and reporters who ever put pen to paper. Their supreme task has always been to create the illusion of reality—to make readers live events. Professor James T. Shotwell observes in his *Introduction to the History of History*:

". . . if the test for the distinction between pre-history and history is the use of writing we may be at another boundary-mark today. Writing, after all, is but a poor makeshift. When one compares the best of writings with what they attempt to record, one sees that this instrument of ours for the reproduction of reality is almost paleolithic in its crudity. It loses even the color and tone of living speech, as speech, in turn, reproduces but part of the psychic and physical complex with which it deals. We can at best sort out a few facts from the moving mass of events and dress them up in the imperfections of our rhetoric, to survive in the fading simulacra in the busy forum of the world. Some day the media in which we work today to preserve the past will be seen in all their inaccuracy and crudity when new implements for mirroring thought, expression and movement will have been acquired. Then, we, too, may be numbered among the prehistoric."

The history of the art of communication is a history of technological effort to achieve immediacy in reporting events. In the chapter that he contributes to this volume Mr. Robert Edmond Jones sapiently dwells on this effort and sapiently regards television as the supreme achievement in attaining immediacy in communication. For it is not the record of the event in which we are primarily interested, it is the event itself, what Mr. Jones aptly calls "the living presence." The telegraph was a huge success in bringing us measurably nearer the event than a physically transported letter ever could. Yet the telegraph must always depend on words or their equivalent in code. It was not until the telephone came that we first experienced a distant event electrically. A conversation between New York and Philadelphia over a wire circuit is itself a living event—something that belongs to the present. It has the immediacy that Mr. Jones demands and the illusion of reality that Professor Shotwell regards as so essential in mirroring thought and action.

To attain this immediacy, this participation in distant events, the engineers have disembodied us. Senses are separated from us and flashed through space—first the sense of hearing and now the sense of sight. All about us are electrical ghosts. They glide over wires across continents. They whisper into our ears. They come and go on television screens. Not ghosts of the dead are they but of living men and women who sit in their homes and their offices and send their discarnate personalities to the uttermost parts of the earth and charge them to deliver messages of love and hate, joy and sorrow, to gather news of life and death, success and failure, and bring it back in the twinkling of an eye. Nothing in the dubious annals of spiritualism approaches this miracle.

CHAPTER ONE

Raising the Television Curtain

BY ALFRED H. MORTON
Vice-President in Charge of Television,
National Broadcasting Company

WE who live in this era, harassed though we may be by swift social change and bitter international strife, are nevertheless fortunate in witnessing the birth of a new social instrument of vast significance—television. For television is here to take its place beside that other amazing commonplace of our daily lives, radio.

I say that we are fortunate, for we are in at the beginning of mankind's greatest venture in mass communication. Television broadcasting, taking its first sturdy steps today, will one day unroll the rich pageantry of a coronation before a viewer in Kansas at the very moment the privileged few see the event at Westminster Abbey. And out along the shores of Puget Sound, in the State of Washington, men and women will be as much participants in the inauguration of a President of the United States as the few who are practically within arm's length of the stand on the steps of the Capitol.

We Present Television

Perhaps it seems to the average man that television still properly belongs among the miracles of the laboratory. So swiftly did the inauguration of television as a public service in America follow on the final press demonstrations in the laboratory, that I am sure that all of us find it hard to believe our eyes when we witness a telecast in some friend's home. We lose sight of the fact that the linking of the essential senses of sight and sound on the air waves is the product, not of a single year's labors, but of more than a half century of profound research and intense experiment. The all-electronic system, freeing television at last from the limitations of whirring belts and spinning disks, was itself built and tested over a period of more than a decade before it was finally given to the public in the United States. Television is ready. And just as the thirties saw radio mature into a potent social force, so will the forties witness the growth and spread of television over the nation.

Television, to be sure, is not yet widespread in its service. The National Broadcasting Company inaugurated its first high-definition service, under technical standards laid down by the radio industry, with the transmission of the opening day ceremonies at the New York World's Fair of 1939. Since that time it has maintained a regular program schedule for pioneer viewers within the area, extending roughly sixty miles in all directions from the Empire State Building, covered by NBC's New York City transmitter. The Don Lee Broadcasting System has also begun a service in Los Angeles under the same technical standards, and the General Electric Company has completed its transmitter near Schenectady for serving the Albany-Troy-Schenectady area. The GE station plans regular relays of New York NBC programs over its service area, thus marking the beginning of the first television network.

America's television services, of course, follow those established in London by the British Broadcasting Corporation by more than two years. It may reasonably be asked why the United States, usually the leader in putting technical advance into practical service, lagged in telecasting. In answering that question, we may well explore some of the troublesome problems that beset American television.

These problems involve vexatious questions of technical standards under which television should operate in a country where radio has grown to greatness and independence under the traditional American policy of free initiative. Here the question is one of agreement within the industry, and approval by the government, for the protection of the public, and ultimately of television itself. They involve, too, questions of extending television over a country of continental proportions, and the great problem of finance. These questions may be stated simply. How shall we do it? How shall we go about spreading it to every John Smith and Jane Doe in America? And, finally, how shall we keep television alive until it is strong and self-supporting?

We firmly believe that television, as it now stands, is ready for the American home. After almost multitudinous demonstrations of older methods of television under laboratory conditions, and some abortive attempts to launch a public service with them, the all-electronic system has triumphed. Almost every telecaster now in the field experimented, at one time or another, with mechanical scanning systems, which are described in the second chapter of this symposium, before abandoning them in favor of the present method. It is apparent that the electronic system is the only one capable of almost limitless improvement without a change in its basic elements.

This very flexibility, however, has given rise to some lively discussion as to what point of technical excellence should be attained before television be considered ready for public participation, and which of several means should be used in realizing some of the various technical standards.

At first glance it would seem to the man in the street that such things as the method and extent of suppressing the lower sideband in transmission, horizontal or vertical polarity, and the number of scanning lines, whether 441 lines or 625 lines, are no concern of his. Yet, of course, this is not true. Television, one of the most complex of all modern inventions, involves the broadcaster's transmitter and the viewer's receiver. And presumably both have made, or would have to make, considerable outlays of money to acquire their respective instruments. In television, as David Sarnoff, President of the Radio Corporation of America, has pointed out, transmitter and receiver must fit as lock and key. A change in the one necessitates a corresponding change in the other. This problem has been largely absent from sound broadcasting; an obsolete receiver will still reproduce the program transmitted from the most modern station.

In television—and I speak here of a public service and not of laboratory experiments—the precise method of transmission and reproduction is all-important to the public. It is rather difficult to explain the reason in layman's terms, but I believe the correct analogy is to be found in railroad history. Early American railroads employed a variety of gauges, so that it frequently happened that the cars of one road could not travel over the rails of another. Television standards in the United States must be uniformly adopted by all telecasters or much the same condition will obtain; the receiver for one station may not reproduce the programs of a competing station in the same city. Neither the public nor the television industry could afford such a luxury.

In England, where broadcasting is supported out of license fees levied on every radio receiver, the British Broadcasting Corporation established a television service at London in November, 1936. Being a monopoly, the BBC settled the troublesome question of technical standards by itself laying down what it considered, after due study, the best practicable specifications. Radio manufacturers thereupon made receivers to conform to those specifications. Thus every make of receiver would reproduce the programs transmitted by BBC's television station at Alexandra Palace.

America faced a quite different problem. The very policy that has made its sound broadcasting of the highest technical quality in the world stood squarely in the way of early inauguration of a television service. Definite and precise standards may be imposed on manufacturers by a broadcasting monopoly, but not by a fiercely competitive group of broadcasters. A committee of engineers representative of the American radio industry, therefore, wrestled with the problem for years before deciding upon a complete group of standards. Many of the devices embodying proposed standards were tested over NBC's transmitter in New York City during the years 1936-38. As a result these standards, which have been submitted for the approval

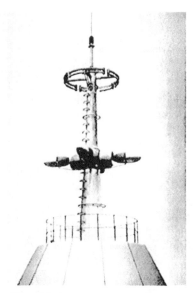

Transmitting antenna array atop the Empire State Building for NBC's New York television station; the upper section transmits sound and the lower, sight.

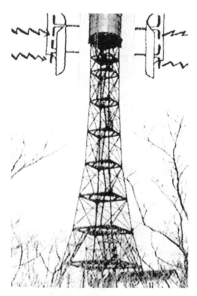

Automatic radio relay tower for transmitting television programs over intercity networks. (COURTESY RCA COMMUNICATIONS)

A typical home television receiver. (COURTESY GENERAL ELECTRIC CO.)

A section of coaxial cable used to convey television signals. (COURTESY BELL TELEPHONE LABORATORIES)

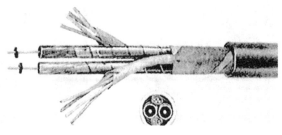

The control room, with its staff of engineers and the program director guiding the technical and artistic quality of the program, overlooks the studio. (COURTESY NBC)

The NBC television studio in Radio City which provides for fluid camera and lighting technique. Three cameras are used simultaneously and each lighting unit is remotely controlled. (COURTESY NBC)

of the Federal Communications Commission, are the highest in the world, a fact which should make them a practicable blueprint for an American service. They provide not only for present needs, but also leave a wide margin for future improvement in the technical quality of the television image. They provided the answer to the first question: How shall we do it? As soon as they were adopted by the committee, NBC made its transmitter conform to the new standards, preparatory to inaugurating a service of regular television programs in the New York City area.

It would be wonderful if one had no need to consider cost in launching television, or even if the end of the developmental years also terminated the unprofitable period. Sad to say, this is not the case. Development of television has cost the Radio Corporation of America alone many millions of dollars, but the investment required before television as a public service reaches its years of self-support will unquestionably dwarf the cost of development. Studios and transmitters must be built and programs must be telecast. And the last is the most expensive item of them all.

We propose to build television as a service along lines laid down by American sound broadcasting, that is, without financial aid from government. Some of its programs, containing commercial announcements, will be sponsored by advertisers. The income derived by the telecaster from such sponsored programs is to support the entire program structure. We have not yet, of course, reached this point.

Our present dilemma was not unexpected. Telecasting is expensive. A transmitter and studios must be built and staffed. Actors must be hired and rehearsed; films must be rented. Stage sets must be built and erected in the studios, and make-up and costumes provided. We would normally, as I have pointed out, expect to pay these costs out of income derived from sponsored programs. Sponsors, however, must find such advertising profitable. Before it can become profitable a large number of viewers must be tuning in on television.

Such an audience is not to be created by waving a wand. Before the prospective televiewer goes to his radio dealer and buys a receiver he must have a satisfactory program service. Which brings us back to the original point of departure, the broadcaster. This sort of dog-chasing-its-tail proposition has been called the "vicious circle." There would seem to be no prospect of an extremely rapid solution to this problem. In fact, the apparent solution lies in stimulating the growth of the dog, so that the circle he traces becomes progressively larger.

We were fully cognizant of the nature of this problem when we began telecasting on April 30, 1939, when President Franklin Delano Roosevelt and others appeared in the opening program relayed by our mobile television station from the Court of Peace, New York World's Fair. Our plan was, as has been indicated, for the broadcaster to bear the expense of a program service in the hope of recouping his outlays at some future time. A minimum service of two hours a week for home viewers was provided for the first few weeks, together with some twenty hours of motion picture transmissions for trade demonstration purposes. Simultaneously, commercial receivers with reproducing tubes of five-inch, nine-inch, and twelve-inch diameters went on sale in New York City radio and department stores. Prices, since drastically reduced, ranged from $199.50 to $600. The program service has since been increased to an average of about fourteen hours a week for home viewers.

Behind the launching of this first American television service lay three years of experimental telecasting, stretching back to July, 1936, when NBC's transmitter went on the air with the all-electronic system yielding an image in 343 lines, at the rate of thirty complete images a second. It was decided after some months of trial that a higher standard, namely, 441 lines, was necessary to provide sufficient entertainment value for the viewer. Accordingly, the 441-line image began going out on the air in 1937.

It was also realized that if television were to become

a public service, then the experimental telecaster would also have to accumulate experience in programming. For, needless to point out here, television may resemble both motion pictures and the legitimate theater, but it has problems and possibilities that set it apart from both. The new art's possibilities and immediate needs were explored in systematic manner during the two years and ten months of experimental transmissions. Each series of telecasts was followed by a period of intense soul-searching and considerable testing for possible flaws in the system itself. As a result, both the quality of the image and the presentation of programs moved forward in orderly and satisfactory fashion.

At the time of inaugurating our regular television service, we had a single television transmitter, one studio designed for the presentation of live talent programs, one studio for the scanning of motion picture film, and a complete mobile television station for relays of such outdoor events as we considered of general public interest. The staff, necessarily kept to minimum proportions during the experimental period, jumped rapidly to about seventy persons, of whom the majority were in the technical division. Some of these men have been connected with the development of television at NBC for a decade and more, following through from the old mechanical scanning days when images, notably those of Felix the

Cat, were transmitted from a studio in the New Amsterdam Theater Building. Program directors entered television at a much later date. In 1937 the staff consisted of a single person, Thomas H. Hutchinson, the present manager of our program department. Six directors made up the staff when our regular service began in 1939 and since that time we have added several more. Stage managers and numerous other individuals indispensable in production have likewise been added to the staff.

With this group—and I assure you it is still of minimum proportions—we are expected to provide a satisfactory service to pioneer televiewers within the limits of a stated budget. Both staff and budget are decidedly limiting factors in the provision of that service. The personnel may be expected to produce only a certain number of live talent shows; the transmitter and studio technical crew can keep the station on the air only a certain number of hours before reaching the end of their respective work weeks. At the present time, therefore, we are telecasting about four hours a week of live talent programs, a like amount of film programs, and about six hours of special event telecasts, including boxing, wrestling, parades and the like, relayed from the field by our mobile television station.

The service described just about exhausts the possibilities of our physical plant and the television staff using it.

We Present Television

The budget, with some very nice figuring, will fill the hours with film and live talent program material, after deduction of salaries and maintenance costs. Any considerable enlargement of the schedule will again require additional funds and, in all probability, more men.

The question of money, it is clear, is a very important factor in television, as indeed, where is it not? Most industries, however, show some kind of income immediately they begin serving the public. The television broadcaster undertakes a service which will probably not return him one cent for several years. On the contrary, he must exercise his judgment with the utmost nicety if he is not to suffer extremely painful financial wounds.

The job of the chief of any television service, in brief, is to see to it that the growing army of televiewers is provided with an interesting and otherwise satisfactory program service. The service must be sufficient in extent to give the greatest stimulation to receiver sales, short of wrecking the financial structure. His director of programs must be given just enough money to enable him to build interesting programs, and no more. And, at the same time, the man in charge must search about for advertisers, or advertising agencies, who can be interested in underwriting part of the cost of one or more television programs in exchange for commercial "plugs."

At first glance all this would seem to be comparatively

simple. But doubling the amount of 'money spent on talent for a given program, for instance, will not yield a commensurate rise in that program's entertainment qualities. In fact, it would probably raise the quality only about 10 per cent, so much is the program dependent on the abilities of the program director. Likewise, the provision of a daily program service of eighteen hours, even were we suddenly to find ourselves up to the task, would not give a corresponding boost to sales of receivers. The only result of such a move would be an invitation to the receiver in bankruptcy and a sudden collapse of all television broadcasting.

Limited as to both scope of operations and budget, the broadcaster finds it of greatest importance that he make every dollar count in filling the hours on the air with interesting programs. He might follow the lead of radio and exercise his own judgment. The imperative dictate of the theater box office is, of course, absent. We know some of the things we can do satisfactorily, and we are only too aware of many things that lie beyond the limits of both our present abilities and our purse. In some large measure we do exercise our judgment, since television, whose age is measured in months, still has no rigid rules. But what of the programs we actually telecast? Are they successful? To measure the degree of success, we instituted a unique poll in October, 1939.

We Present Television

The method is simplicity itself. We invite, during our television programs, each member of our audience who owns a receiver to send in his name and address for addition to a list of such owners. In return we send him a weekly schedule of programs. Attached to each issue of the schedule is a return postal card, bearing spaces for the rating of each program item telecast by NBC. The ratings are "excellent," "good," "fair," and "poor." To these ratings we assign numerical values of 3, 2, 1, and 0, respectively. The average values, computed on the returns, give us a clear indication of the entertainment or educational value of each program.

Every television set owner in the audience, therefore, has an opportunity to voice his opinion as to the quality and acceptability of each individual program. We feel that in offering this opportunity to the audience, and keeping our thoughts and operation flexible, we can readily trim our programs as closely and as quickly as possible to the majority vote of the audience. By the closeness and frequency of our contacts with viewers they have become our partners in launching and studying this great new force.

The poll, which at the time of writing had increased to more than two thousand names, embraces about one-half the present receiver owners in the area served by the NBC station. Its current rate of increase is about

4 per cent weekly. It is interesting to note that not only has the percentage of returns been quite consistent, but also that the high average of about 40 per cent return is quite unusual for a poll of this nature.

Results of the systematic polling of our audience so far have borne out some of our previous beliefs. In other respects, however, we find that we were either slightly or greatly mistaken on individual programs. In addition to rating each individual program, the poll also gives us a satisfactory measure of the general level of all our program efforts.

I have thus far only indicated the nature of some of our problems. I hope that I have not conveyed the idea that the general problem of television is, in any single respect, static. The television audience will increase and multiply; more money, more men, and more hours on the air will be needed, but coincidentally will come opportunities for earning, at first a part, then all, of the expenses involved in telecasting. Naturally, the telecaster will do everything he can to speed that day.

We have found advertisers and advertising agencies keenly interested in television. Some of them, excluded by the nature of their goods, have had no success in sound broadcasting. Others, more or less successful in radio appeals, are aware that television will open new opportunities for promoting the sale of their goods. In increasing numbers advertisers have brought in programs for telecasting, eager to discover for themselves some of the potentialities of television advertising. Undoubtedly the volume of these experiments will continue to mount until one day an advertiser will decide that the moment has come to sign a thirteen-week contract for time on the air.

Just how far in the future that time will be depends on a number of factors. Out of his own pocket, as I have said, the telecaster must provide a service satisfactory as to both quality and quantity of programs for maximum stimulus to receiver sales. Here is the answer to the question: How shall we keep television alive until it is strong and self-supporting? On the other hand, the manufacturer must build receivers that are both reliable and within the purse limits of the average American. With these two problems properly met, commercial television should become a reality within a few short years, thus supplying the ultimate answer to the question of television's means of support.

A great aid in the solution of most of the telecaster's problems will come with the beginning of network television. Broad hints at the comparatively high cost of television programming, as compared with radio, have been scattered here and there in this contribution. And while we who have undertaken to provide a program service over a single television station have no complaint to make on this score, having gone into telecasting with our eyes open, we do look forward to the day when some method of syndication will bring other telecasters in to share the costs of production of a progressively better and more extensive program service.

The problem of syndicating radio programs was largely solved by the time network companies came into being. The intercity lines of the American Telegraph & Telephone Company were already in existence when networks began stretching out fanwise from New York City. And although many improvements have been incorporated in the system, it is still basically the same system as that which existed at the beginning of network radio. The same means, however, are not available to television.

Several methods of syndication have been suggested. The one most prominently mentioned over a period of years is the coaxial cable, a tubular line conductor. This cable, however, is extremely expensive. An experimental installation, linking New York City and Philadelphia, was made several years ago at an estimated cost of approximately $5,000 a mile. That price is too high for television.

Lately a new device has been brought out of the laboratory. The Radio Corporation of America staff, after several years of research and experiment, have announced that an automatic radio relay is ready for television's use. The rather formidable name covers a smallish device which is at once a receiver and a transmitter, and operates without the attention of a human hand. Mounted on steel towers one hundred feet high and spaced at intervals of from twenty miles to about fifty miles, a line of these relays would be used in place of the considerably more expensive coaxial cable.

Regardless of what method is used—and perhaps a combination of cable and radio relay may be the ultimate means of linking stations together in a network—it is fairly certain that in less than three years a regional network, stemming from New York City, will stretch along the Atlantic seaboard from Boston to Washington. It will serve a thickly populated area, containing about one-fourth of the nation's people. Such a network, accompanied by a progressive merchandising policy on the part of radio manufacturers, should be fairly tempting to the advertiser.

A second regional network should spread somewhat radially from Chicago, and a third proceed northward from Los Angeles. All of these networks should be in existence within a period of five years, perhaps less, with the linking of all three into a nation-wide span scheduled for some years later. The National Broadcasting Company has filed applications with the Federal Communications Commission to build additional television stations at Philadelphia, Washington, and Chicago. These cities may be connected with New York with the automatic relay transmitters already mentioned. It seems very much like dealing in remote futures to speak of these things, but from experience dating back into the years before 1939, when all of us stood pretty much in awe of television and wondered whether the day would come when it would be in public use, I can testify that such eventualities come to pass very quickly.

In this contribution I have touched lightly on some of the very heavy problems that hang over the American television broadcaster. There are, of course, a multitude of others, and I do not doubt that they will increase in number before they become fewer. My one hope is that these paragraphs have brought the reader to some understanding of the magnitude of the struggle that is beginning to carry television into the American home. Heavy though the labors may be—and I can assure the reader that they are heavy—I do not believe that anyone who is connected with the nascent art of telecasting would exchange his lot for an easier one in some other activity. The thrill of television is in watching each successive program march a little further toward the future, of seeing the men and women behind the camera do their jobs each week with a little more assurance, and of always and forever anticipating some new development in programming that will definitely mark a milestone in the creation of a new and distinctive art form.

It would hardly be fitting to end this contribution without some consideration of the relationship of television to other media of entertainment and education. It has been suggested rather freely that television must inevitably supersede radio. And on many occasions I have heard it said that a titanic clash between motion pictures and television is likewise inevitable.

One cannot be certain of the future relationship of radio and television. Just as the talkies grew out of the silents in motion pictures, so television is a natural outgrowth of radio. It is, in fact, the addition of sight to sound broadcasting. In some quarters it is held that a broadcast without sight will be as obsolete, within fifteen years, as silent movies are today. Such a development is not inevitable. There will always be a place for music, for instance, without benefit of a view of the musician. Such a program, and other types as well, would serve as a pleasant background for household work, reading, or conversation. But whatever the future holds in store for both television and radio, I am sure that the broadcasting industry will find a way to reconcile the two in a satisfactory manner.

Raising the Television Curtain

I feel that the opposing of television and motion pictures, in most persons' minds, comes about as a result of misunderstanding. Television makes its appeal to members of the family seeking relaxation in the home at some point during the day or evening. Its entertainment and educational appeals must meet this psychological want. Motion pictures and all other forms of the theater make a quite different appeal. Usually the movie patron goes to the cinema because he wants to break away from the routine of home or office, or both. Going to the movies is an event. The housewife wants to leave familiar walls to mingle among others on similar bent. She, and probably her family too, dresses for the event and is quickly transported from the cares of the day in the glamour to be found in the motion picture. For this reason I find it impossible to believe that television will ever seriously challenge the existence of motion pictures.

On the other hand, I find every reason why motion pictures and television should find mutual profit in close co-operation. Hollywood has film, and television needs film. From the surplus footage shot in the making of a Hollywood feature, television could make many an interesting hour of program, without jeopardizing the feature's chances of success at the box office. Likewise Hollywood's surplus of men and equipment could be put to work creating stories on film expressly for television.

We Present Television

We of television are anxious to come to an understanding with the motion picture industry, and I believe that within a short time Hollywood will see an added source of profit in co-operating with television.

It is yet too early to make any prophecy as to television's future character. Who, for instance, could have foretold accurately, after watching the flickering movies of the nineties, or witnessing the production of a crude film in the studios of Edison or Méliès, what Hollywood would be like today? But we can all be certain that in television mankind has created a new and mighty social instrument. Television is great. It will become far greater. And as to its value in future society, I cannot do better than quote David Sarnoff. "The ultimate contribution of television," he has written, "will be its service towards unification of the life of the nation, and at the same time the greater development of the life of the individual. We who have labored in the creation of this promising new instrumentality are proud to launch it upon its way, and hope that through its proper use America will rise to new heights as a nation of free people and high ideals."

CHAPTER TWO

The Technique of Television

BY DONALD G. FINK
Managing Editor, Electronics

THE clue to the technique of television lies in one very simple statement: television pictures must be sent on the installment plan. When we look at a scene directly, we see it all at once and we see it continuously. Not so in television. The scenes in television must be sent piecemeal, bit by bit, and each piece of the scene must be reassembled before the eye at the receiver. Moreover, the scene must be reassembled very quickly, so that the eye is not aware of the process.

In addition, to allow scenes in motion to be televised, the scene must be "photographed" many times per second; that is, a great many separate and complete pictures must be sent, one after the other in each second. In this respect television is similar to motion pictures, since both employ the artifice of presenting to the eye a rapid succession of still pictures, each differing slightly from the preceding and following ones. In this manner, the mo-

229

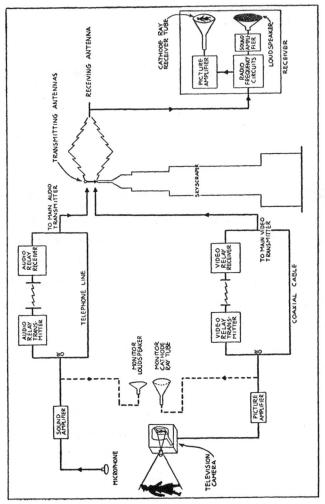

Simplified diagram of complete television system.

tion in the scene is broken down into a series of smaller motions which may be photographed without blurring. When the succession of pictures is presented to the eye rapidly enough, the screen seems to be continuously illuminated by the scene, and the motion appears smooth and continuous. The only essential difference between the movies and television is the rate at which the separate pictures are presented: 24 pictures per second in the movies and 30 pictures per second in television, according to present standard practice. This discrepancy, incidentally, makes it rather difficult (but still quite possible) to transmit standard motion picture film programs over the television system. The movies have adopted the rate of 24 "frames" (as each picture is called) per second for the very good reason that this rate offers a good compromise between the cost of film on the one hand and the satisfactory presentation of the subject matter on the other. When television came along, 30 frames per second was deemed an advisable figure because of the type of alternating current usually used to operate television receivers. The difference between the two rates has been satisfactorily overcome by the special movie projectors used in television studios. Essentially, so far as picture repetition goes, motion pictures and television use the same technique.

We Present Television

But here the resemblance ends. In the movies each separate frame, or picture, is presented to the audience in its entirety, all at once. In television, on the other hand, each separate picture must be broken up into several hundred thousand tiny points of light, and these points of light must be conveyed to the receiver one at a time. This is the installment plan with a vengeance. The requirement of dissecting each picture into its essential elements, tiny points of light, is the reason why television is such a difficult art technically. It explains why television took so long in reaching a form suitable for the public, as well as the reasons why the new art faces so many restrictions.

The necessity for this piecemeal method of transmitting television pictures lies in a simple but inescapable property of electrical communication systems. A single electrical communication circuit can transmit but one item of information at a time—no more. In the human eye—which is in itself an electrochemical communication system—this limitation is circumvented by providing many hundreds of thousands of separate fibers in the optic nerve, each of which operates simultaneously with all the others. The eye is thus enabled to see the whole contents of a scene at one glance. But in television we cannot employ hundreds of thousands of communication circuits at once. In fact, we must be content with just one link, that existing between the broadcasting station and the receiver. This one link is called upon to handle the whole television process, 30 pictures per second, each picture dissected into roughly 200,000 tiny points of light. These two figures multiply to the startling figure of 6,000,000 points of light to be conveyed, successively, over the system, each second. This extremely rapid rate of conveying information makes the modern television system by far the most comprehensive means of electrical communication yet in existence.

For comparison, consider the ordinary telephone circuits used to connect the stations of the national networks for sound broadcasting. These circuits are capable of handling approximately 5,000 items of information per second (here an item of information is an electrical vibration corresponding to the sound wave which is picked up by the microphone). Such a circuit can handle speech and music with a very considerable degree of realism. The television system, in contrast, must work at a rate some 1,000 times as fast. The ancient Chinese have said that a picture is worth ten thousand words, which is not such a bad guess.

Before going on to examine how the television system performs its difficult task, let us examine in more detail what we mean by the tiny dots of light which the system must convey from transmitter to receiver. In the illustration facing page 49 is shown an ordinary photoengraving (upper left corner) and a three-times enlargement which reveals the structure of the reproduction. The enlargement shows the picture to be made up of a great many printed dots. In the darker portions of the picture the dots are of large diameter, forming an almost solid black mass in the black regions. In the brighter portions of the picture the dots are of small diameter and in the high lights they may be missing altogether. By this method of printing, the amount of light reflected from the page varies in accordance with the contents of the scene. In television a very similar process is used. The tiny points of light previously mentioned correspond with the half-tone dots in the engraving. In television transmission, an electrical impulse for each one of these dots must be sent over the system, one after another, fast enough to allow the whole picture to be reassembled before the eye in $\frac{1}{30}$ of a second.

The illustration shows the reason why television images must be divided into so many points of light. The enlargement is obviously too coarse to allow a satisfactory portrayal even of a close-up, and if a distant scene were in view, the coarse structure would obscure the contents of the scene altogether. Suppose, however, that the whole area of the illustration were covered with detail as fine as that shown in the upper corner. Then a much more satisfactory reproduction is possible. The enlargement contains (count them if you wish) no less than 15,000 halftone dots. If the whole picture contained detail as fine as that in the upper corner, the scene would contain about 130,000 dots, which is close to the 200,000 figure previously mentioned. In other words, the modern television system, working at its best, should be able to do somewhat better than the reproduction shown in the corner, when reproducing a picture the size of Figure 1.

It should be remarked that this degree of detail, while satisfactory for most purposes in television programs, compares very poorly with other means of pictorial reproduction. For example, a fine-grain photograph, printed by contact, contains some 50,000,000 different points of light in the space of an 8-by-10-inch print. The average 35-mm. professional movie, as presented to the audience under average conditions, contains about 500,000 points of light, the average 16-mm. home movie about 125,000. This fact accounts for the oft-quoted assertion that the detail of television pictures is about the same as that of a good home movie taken on 16-mm. film. The performance of the television system is definitely superior, so far as pictorial detail goes, to the 8-mm. home movie.

We Present Television

From Camera Tube to Picture Tube

So much for the questions of pictorial detail, and how much of it can be handled by the present television system. Let us turn now to a brief description of the equipment used to perform the television miracle. We begin with the camera, which views the scene to be televised in much the same manner as does the ordinary motion picture camera.

The television camera is essentially a closed box fitted with a lens which admits light to the interior and focuses the scene on the sensitive plate within. The sensitive plate is placed in the same position as the film in an ordinary camera, but it operates in quite a different way. Ordinary photographic film is affected by the light in a photochemical way, that is, the image is brought out and reproduced by chemical means. In the television camera the image is developed photoelectrically, that is, an electrical process takes place on the sensitive plate.

The television camera differs from the photographic camera in another way: In ordinary photography the whole picture is taken at once. In the television camera the whole scene is received at once, but it must be dissected immediately into its constituent points of light. Thus the camera must perform three functions: it must have a lens to focus the image; it must be capable of transforming the light image into an "electrical" image; and it must dissect or "scan" the image. The lens performs the first function. The second two functions are performed by the camera "tube," which is enclosed within the camera housing.

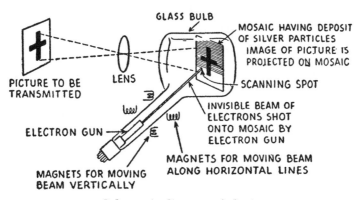

FIGURE 1. *Schematic diagram of the iconoscope.*

There are several forms of camera tubes now available for television use, but the one most widely used, and the most significant in the modern history of the medium, is the type known as the iconoscope. This tube (see Figure 1) is a dipper-shaped glass structure from which the air has been exhausted and inside which is the sensitive plate previously mentioned. The iconoscope is so placed in the camera that the sensitive plate lies opposite the lens, and thus receives the image.

We Present Television

The sensitive plate itself is a thin sheet of mica on the front surface of which have been deposited several million tiny droplets or globules of silver treated with a surface layer of cesium and oxygen. When light falls on these silver globules, the silver gives off minute particles of negative electricity (electrons). The electrons given off are attracted by a positive charge placed on another electrode in the tube, and are thus removed from the sensitive plate. Moreover, the amount of charge thus lost by each globule depends directly on the amount of light falling on that globule.

Thus, all over the surface of the sensitive plate, negative electricity is released by the action of the tiny points of light which make up the image of the scene. The result is that the sensitive plate acquires a deficiency of electric charge which corresponds, point for point, with the lights and shadows of the scene to be televised. The longer the image of the scene continues to fall on the plate the greater the deficiency becomes, and in this way the iconoscope stores the light until it is ready for use. The important function of the sensitive plate is, therefore, to transform the optical image into a corresponding electrical image and to preserve the electrical image until it can be dissected into its elemental points of electricity which correspond with the elemental points of light in the original scene.

The Technique of Television

The third function of the camera, dissection of the image, is performed by an "electron gun" located in the side arm of the tube. This device sprays a stream (or "beam") of electrons, much like water from the nozzle of a hose, at the sensitive plate. The stream of electrons is very sharply defined, and it may be directed at any point of the plate by a pair of magnetic coils arranged around the arm of the tube. The current passed through these coils is used to direct the electron stream over the surface of the plate in a series of horizontal lines, back and forth, until the plate has been "scanned" from top to bottom. The motion of the electron stream is very similar to that of the eye in reading a page of printed matter. The beam moves across the top of the picture, then quickly reverses its motion and begins again at the next line, and so on until it has traversed the plate over its entire surface.*

The electron stream, in other words, explores the sensitive plate systematically, to determine the information contained on it. Whenever the electron stream hits one of the silver globules, it restores the charge deficiency on the globule to equilibrium, and the amount of charge restored is, of course, determined by the amount previ-

* Actually the beam traverses the plate in two series of lines alternately, one set of lines filling in the spaces between the set previously traced. This technique, known as "interlacing," is employed to avoid flicker in the reproduced pictures.

ously lost because of the light falling on the plate. Thus, as the electron beam moves successively over the silver globules, it restores to each an amount of charge corresponding to the amount of light falling on that globule. The charge restorations in turn give rise to voltage impulses which are conveyed from the camera tube to the television transmission circuit. Thus the camera tube generates a succession of voltage impulses which correspond to the values of light and shade successively passed over by the electron stream. The two-dimensional picture is thereby transformed to a succession of electrical impulses which contain all the information inherent in the scene.

This dissection process, as we have seen, must take place at a very rapid rate. The picture must be dissected or scanned in $\frac{1}{30}$ of a second. Furthermore, in order to convey sufficient detail, the picture must be divided into about 500 lines (the present standard is 525 lines from the beginning of one picture to the beginning of the next). This means that some 15,750 lines must be covered by the electron stream, in 30 successive pictures, during each second of the performance. Each line in the picture contains about 500 points of light and shade, so the 500 lines contain in all some 250,000 pictorial elements.

More astonishing is the rate at which the electron

stream must move. The sensitive plate in the iconoscope is about 4 inches wide, and the electron stream covers this distance from left to right in roughly $\frac{1}{15,000}$ of a second, that is, at a rate of about one mile per second. The stream returns from right to left at a speed about 7 times as fast, or 7 miles per second. Such a rate of movement is well-nigh impossible in any mechanical system. It is easily possible, however, when the agile electron is used as the agent for dissecting the picture.

The succession of electrical impulses generated by the camera tube must then be conveyed to all the receivers within range of the transmitter. This is done, first, by amplifying the minute electrical impulses, after they leave the camera, by roughly 1,000,000 times, and then imposing the strengthened impulses on a radio carrier wave which is radiated from the antenna.

When the radio wave encounters the receiving aerial, it has been greatly weakened by its flight through space. Hence, after it is conducted to the receiver, it must be amplified several thousand times before the electrical impulses representing the picture may be separated from the radio wave. Once separated, the impulses are again amplified, this time only 10 to 20 times, and they are then ready to control the picture tube.

The picture tube, as shown in Figure 2, is a funnel-shaped glass structure, pumped free of air and closed at the wide end. In the narrow end is an electron gun, very similar to that used in the iconoscope. This electron gun generates a stream of electrons which travels through the tube to the wide end. There the stream encounters a chemical substance which is coated on the inside of the glass face. This substance (usually a compound of zinc,

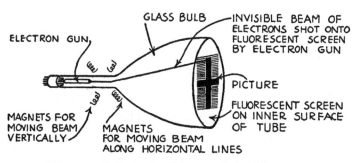

FIGURE 2. *Schematic diagram of the kinescope.*

oxygen, and silicon or sulphur) has the peculiar property of fluorescing, or producing light, when bombarded by the electrons in the stream. In consequence, when the stream hits the screen, a spot of light is formed. The color of the light in modern television tubes is white, although in some tubes it may have a greenish or bluish hue.

The electron stream in the picture tube, as in the camera tube, may be directed at any point on the fluorescent screen, by means of electromagnetic coils (or by means of electrostatic metal plates sealed within the tube). By means of the electric current applied to these coils, the electron stream is caused to traverse the screen in a series of adjacent horizontal lines which match exactly the lines traced out by the electron stream in the camera tube. In this way the surface of the fluorescent material is caused to glow successively along each of these lines, and the motion of the electron stream is so rapid that the screen appears to be illuminated continuously over the rectangular area (see Figure 2) on the screen. This rectangle of light is the illuminated screen on which the reproduced picture appears. It should be borne in mind that, while the screen appears to be illuminated continuously, actually all the light present at any instant is the light present in the rapidly moving spot of light. It is only the persistence of vision which gives the impression of continuous illumination over the whole picture area.

How, then, is the picture reproduced? Here the chain of amplified electrical impulses comes into play. These electrical impulses are conducted to the electron gun, and there they control the number of electrons which enter into the stream. As the number of electrons changes, so does the brilliance of the fluorescent spot. In fact the spot of light may be entirely extinguished, or made to have any value of brilliance up to the maximum of which the tube is capable, simply by varying the control voltage applied to the electron gun. Thus, as the electron stream moves across each line in the scanning motion, the brilliance of the light it produces is varied in accordance with the brilliance of the light in the corresponding line scanned in the camera tube. In this way the picture is reproduced, line by line.

It is necessary, of course, that the motion of the electron stream in the picture tube at the receiver correspond exactly with the motion of the electron stream in the camera tube at the broadcast station. Otherwise the camera tube might be "scanning" a line at the bottom of the picture while the picture tube was scanning a line at the top, and the reproduced picture would have its top where the bottom should be. Or, even if the same lines were scanned at the same time, unless the positions of the two electron streams were identical at every instant, the points of light in the reproduced image would be out of position. So it is essential that the two scanning motions be exactly synchronized. In practice this requirement is met by sending special "synchronizing" signals from the transmitter to the receiver. These signals are generated in equipment maintained at each broadcast station. The signals are used to control the camera tube electron stream, and at the same time they are sent over the air, on the same wave which carries the picture information, so that they are available for controlling the electron stream at the picture tube. Adjustment of the receiver to make proper use of the synchronizing signals is of course highly essential. Most modern receivers have control knobs especially intended to restore synchronization in case the picture reproduction falls out of step with the picture dissection process at the transmitter.

No mention has been made thus far of the synchronized sound system which provides the sound portion of a television program. In general, the sound transmission system is similar to that used in standard sound broadcasting, except that the wave length of the radio waves employed is much shorter. Usually an entirely separate sound transmission system is employed, including the microphone, telephone circuits, and radio transmitter. At the receiver, the radio waves carrying the picture and the sound are commonly received on a single antenna and may be handled together in one or two amplifying stages. But thereafter they are separated, the sight portion being amplified further before controlling the picture tube, and the sound portion being similarly amplified before controlling the loud-speaker. Synchronism between the sight and sound aspects of the program is maintained without any special precautions, because the sight and sound transmission processes are substantially instantaneous. Moreover, whatever small delay (usually only a few millionths of a second) does occur, applies equally to the sight and sound transmissions. In this respect television has a decided advantage over the motion pictures, in which elaborate steps must be taken to synchronize the sound track with the projection of the picture.

Why Has Television Taken So Long in Attaining a Practical Form?

The modern television system we have just examined is a truly remarkable achievement in science and engineering. In fact, the system makes more comprehensive use of electrical technology than any other branch of electrical engineering. Equally remarkable, however, is the length of time which has been required to bring the system to its present state of utility. As J. C. Wilson * has pointed out, television as an idea dates from the inception of the telephone, in about 1875, and practically all of the knowledge necessary for the fruition of the idea has been available since the days prior to the World War of 1914-18. But the particular amalgamation of theory and practice required to produce the practical television system of today has required a very steady effort over the past twenty-five years.

* J. C. Wilson, *Television Engineering*, Pitman and Sons, Ltd., London, 1937.

The Technique of Television

The fact that television systems began to appear, as ideas, so early in electrical history resulted from the discovery in 1873 by a British telegrapher named May that the chemical element selenium possessed the property of transforming changes in light to corresponding changes in electricity. This is obviously a completely essential property to any system of television, and its discovery at once started a great many inventors devising means of applying it in a practical television system. One of the first was that of the Frenchman Senlecq, who proposed in 1877 that a mosaic be built up of selenium cells, each cell to be connected by a separate circuit to a shutter which would drop down when its cell was illuminated. Behind the shutters was a source of light which shone through when the shutters were operated. This system was quite capable of reproducing crude silhouettes, but there is no evidence that it was ever actually built and made to operate. This is, so far as is known, the first proposal for a television system, and for it Senlecq deserves the title "Father of Television."

Following Senlecq's suggestion, a great many others were advanced from 1877 to 1884, all very similar in principle. In 1884 came the invention by the German Nipkow of the rotating scanning disk, which until 1930 was very widely used in television development. Nipkow's disk made use of the very significant technique, previously suggested by several others, of examining the scene to be transmitted and dissecting it into points of light which could be conveyed successively over a single electrical circuit. Nipkow's work ranks high in the history of the medium, principally because he realized so early a system which was not improved upon, basically, for nearly fifty years. Shortly thereafter, in 1890, the Englishman Sutton proposed a system for a television receiver which ranks in importance with Nipkow's system for the transmitter. Sutton's apparatus used a scanning disk and an electrically controlled light source known as a "Kerr Cell." This method of reassembling the image was likewise remarkable in that it was used widely in practical television systems for nearly forty years. Neither of these systems is used to any extent in modern television work, but they were necessary antecedents to more modern practice.

Most of the early work was based on the electrical sensitivity of selenium to light, and this effect was not ideally suited to the purpose. In the first place, selenium cells are rather slow in action and hence incapable of following rapid changes in light which occur in images in motion. In the second place, the selenium cells were highly variable in their sensitivity and far from permanent. So it remained for the early television workers to discover the utility of the true photoelectric effect, which

is the basis of the light-sensitive action of the silver globules in the iconoscope. The photoelectric effect was discovered by Hertz, the "Father of Radio," in 1887, when
he noticed that a spark could be made to jump over a
gap much more readily if one of the electrodes were
illuminated with strong light than if the event occurred
in darkness. The German Hallwachs later studied the
effect systematically and came to the conclusion that the
light set free electrical particles from the surface of the
electrode. Sir J. J. Thompson then identified the particles
as electrons, and Albert Einstein came forth with his
theory of the photoelectric effect, on which his fame as
a physicist rests as much as on his theory of relativity.

The practical side of the photoelectric effect was advanced by Elster and Geitel, who as early as 1890 built
practical photoelectric cells which would pass electrical
current when light fell upon them, but not when they
were in darkness. These cells, the precursors of the modern "electric eye," were very important tools in the hands
of the television worker, and have since proved, in one
form or another, to be the backbone of the television
camera.

The specifications for the modern electronic system of
television described in the beginning of this chapter were
laid down, surprisingly enough, in 1907 by two workers
independently. The Russian Boris Rosing and the Eng-

lishman A. A. Campbell-Swinton came forth in that year with the proposal for a light-storage camera tube very similar in principle to the modern iconoscope, and suggested at the same time that the picture-reproducing apparatus be a "cathode-ray tube" very similar to the electronic picture tube of the present day. The cathode-ray tube had previously been developed by Braun as a means of studying the behavior of electron streams. It has since been highly developed as an electrical tool, but its practical use in a television receiver was to wait until 1928.

The development of the light-storage picture tube was equally long in completion. In fact, the first successful attempt to put Rosing's and Campbell-Swinton's ideas into practice came in 1928, when V. K. Zworykin applied for his patent on the iconoscope. The iconoscope has been vastly improved since that time, but credit for the first camera tube capable of transmitting high-definition television images surely belongs to Zworykin. This important invention is one of the first American contributions of basic importance.

Another American worker of great importance in television is Philo T. Farnsworth, who has invented several forms of camera tubes, of which the "image-dissector" is perhaps the most widely used (for film transmissions primarily, since it does not have sufficient sensitivity to pick up studio or outdoor scenes unless the lighting is very intense). Farnsworth and his associates also have contributed much to the problem of deflecting the electron streams used in electronic camera tubes and picture tubes.

On the practical front of television system development must also be mentioned the American C. F. Jenkins and the Englishman John L. Baird. Baird is given credit for transmitting the first half-tone television images in motion, in 1925, although Jenkins achieved the same result within a few months. These images were extremely crude, but they were improved to form the basis of the early systems of mechanical television which were shortly thereafter made available to the public. Several broadcast stations here and in Europe broadcast regular programs to the public from 1928 on, and simple receivers were available during the ensuing five years. However, the pictures possible in the mechanical system of scanning did not have sufficient pictorial detail to support an entertainment service, and public interest lagged.

Thereafter it became clear that electronic methods were essential, and they were developed with vigor by several commercial engineering concerns. One of the first organizations to offer a public program service based on electronic methods was W6XAO of the Don Lee Broadcasting System in Los Angeles. This station began transmitting scheduled programs in 1933 and has been on a regular schedule ever since.

In 1936, the National Broadcasting Company and the Radio Corporation of America began a series of field tests in New York City, using the all-electronic system. These transmissions were not intended for the public, but were viewed by members of the technical and executive personnel of the two companies. On April 30, 1939, these transmissions were extended on a regular schedule for reception by the public. Coincidentally several manufacturers of television receivers offered models for public sale. This event was the first co-ordinated effort to establish television service for the nontechnical public (the Don Lee transmissions on the West Coast had been received for the most part on receivers built by amateurs, since few commercially manufactured receivers were available). For this reason the opening of the New York World's Fair, on April 30, 1939, is commonly viewed as the first major telecast for the public in the United States.

The Boundaries of Television—What Can Be Done and What Cannot

The modern television system has great capabilities, and equally great limitations. We have already discussed one phase of the area within which television can operate, namely, the pictorial detail of its reproductions. The present television system is capable of reproducing a picture about the equal of 16-mm. home movies (although not every broadcast, nor every receiver, attains this level in the present state of the technique). This degree of detail is an achievement, but it is also a limitation which the program producers must keep in mind in the programs they offer. The detail available is perfectly adequate for close-up pictures, but it becomes less so as the figures on the screen become smaller, that is, as the camera is removed further from the object. Suffice it here to say that by the proper use of cameras with interchangeable lenses, it is possible to cover an area as large as a football field without causing the audience to be aware of the detail limitations, but it can be done only with expert use of the cameras.

Another limitation of less pressing importance is the fact that the picture is reproduced in black and white, rather than in colors. The universal experience with motion pictures has so accustomed us to this type of reproduction that we rarely question it, but it is important nevertheless. Scenes in nature, viewed in full color, display contrasts of color which are partially, if not wholly, lost in the black and white reproduction. Part of this lost contrast may be restored, but to do so it is important that all the colors in the rainbow receive equal treatment from the television camera. The early iconoscopes displayed a sharp preference for the red and infrared—so much so, in fact, that the silk lapels of dinner jackets often televised white or gray because they reflected so much infrared light. Recent improvements in the iconoscope (as well as in other types of camera tube) have removed this oversensitiveness to red. In fact, the modern iconoscope is capable of dealing with colored subjects substantially as well as the panchromatic films now used in motion picture production. From this point of view, the color performance of the modern television system may be said to be completely satisfactory.

A third item of importance in the operation of television cameras is their sensitivity to light. How bright must a scene be before it can be properly televised? The best answer to this question can be stated in terms of the sensitivity of an equivalent photographic film. At present the most sensitive camera tube is the orthiconoscope ("orthicon" for short), an improved version of the iconoscope. The iconoscope, previously described, produces undesired signals, or "dark spots," in the picture, by the process of scanning. The orthicon (shown in Figure 3), however, does not develop these surplus signals, and the operations needed to remove them in the iconoscope are not necessary. The orthicon has a sensitivity which approaches that of the films used in motion pictures, which means that it can televise a scene in ordinary room illumination. For the best results, however, more light is desirable, and in studio practice it is customary to employ several hundred foot-candles of light.

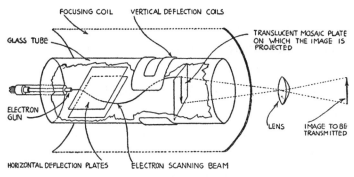

FIGURE 3. *Schematic diagram of the orthicon.*

This is a large amount of artificial illumination, and it is accompanied by a considerable amount of radiated heat. The heat is one of the commonest complaints of television performers. Although the light is no brighter than that employed in motion picture studios, the duration of each scene in television is often longer than in the movies, and hence the over-all effect of the heat is greater.

In contrast to such high illumination, however, can be cited the experience of the NBC crews in televising the end of a football game on a cloudy day in the fall of

We Present Television

1939. On one such occasion the light was so dim that it produced no deflection on the standard Weston photographic exposure meter, but nevertheless a recognizable television image could be transmitted. The sensitivity of the television camera is being rapidly improved at present, and it is not too much to suppose that eventually it will be able to work under any light conditions in which a movie camera may be used, if indeed it does not surpass the motion picture camera.

One of the most widely heard questions about television reception is that concerning the small size of the picture. Receivers commercially available at present produce pictures ranging from 3 by 4 inches to about 8 by 10½ inches. These picture sizes are small compared with the screens used in home movies, for example, but they are nevertheless quite capable of giving satisfaction, provided the audience sits close enough to the screen. The important factor, so far as picture size is concerned, is the distance at which the picture is viewed, relative to the picture height. The usual preference of audiences, whether the picture is large or small, is to sit at a distance about 5 times as great as the picture is high. For a picture 8 inches high (the maximum in commercial receivers) the corresponding viewing distance is somewhat more than 3 feet. At such a viewing distance the picture does not seem small, but obviously only a few people can sit as close to the screen as this. Television pictures as large as 18 by 24 inches, with a viewing distance of 7½ feet, are desirable for the usual family audience in a living room. That such pictures will become available in the future cannot be doubted. In fact, television pictures as large as 12 by 15 feet have been successfully demonstrated in theaters, but the necessary equipment is not suitable for the home because of its high cost and technical complexity. Cathode-ray picture tubes used in home television receivers have been built to accommodate pictures as large as 12 by 16 inches, but this is the upper practical limit at present because of the expense of the tube and its associated circuits.

Finally, one of the most important boundaries of present-day television is the distance over which the programs may be transmitted. Because of the tremendous rate at which information must be sent over the television system, a very large amount of space must be used in the ether by each television broadcast station. Actually the space occupied by one television station could accommodate 600 standard sound broadcast stations. Accordingly, it has been necessary to assign television stations to portions of the ether spectrum where space for them could be found, that is, in the short wave length region. Even in this region, room has been found for only 19 television station channels, and only 7 of these are considered useful at present for public service. There will be available, therefore, only 7 choices of program in any given locality, when a sufficient number of stations have taken the air. This does not mean that 7 stations can be assigned in any one city, but they can be assigned within any area of, say, 200-mile radius, beyond which stations do not ordinarily cause interference with other stations.

By far the most important result of the use of very short waves for television transmissions is the limitation of distance. It is a good working rule (but not without exceptions) that the distance over which such ultra-short waves may be depended upon to give a reasonable quality of television reception, is limited to the horizon as viewed from the transmitting antenna. The distance to the horizon increases with height, so it is desirable to erect the transmitting aerial on as lofty a pinnacle as possible. In New York City the two tallest skyscrapers, the Empire State Building and the Chrysler Building, have already been pre-empted for television stations. In the case of the Empire State Building the horizon is located about 45 miles away, and consistent reception seems to be limited to about 50 miles except in favorable receiving locations. For the more usual case in a smaller city, the aerial may be raised no higher than about 250 feet, and in this case the horizon is 25 miles. Fortunately, the relation between the coverage of a station and the size of the community is such that most urban populations can be covered by stations within the city limits. But rural areas, as well as some suburban areas, cannot depend on television reception, if indeed it can be made available to them at all. So for the present, at least, television is an affair for the cities and the clusters of communities around cities. Fortunately for television, the bulk of the nation's population resides in these areas.

At the receiving end, the location of the antenna is equally important. The antenna should be mounted as high in the air as possible and free of obstructions. Moreover an ordinary piece of wire, which serves well for most sound radio installations, is not suitable for television. Rather a specialized structure, known as a "dipole aerial," is desirable because such an aerial is sensitive and because it is comparatively free from signal reflections which cause "ghost" images in the received pictures. The dipole is simple: it consists of two rods, each four to five feet long, mounted horizontally on a wooden or metal mast, so that the whole structure resembles a "T." The two rods are placed at right angles to the direction from which the broadcast comes. The lead-in wire is composed of two conductors, one connected to each of the two rods.

When the receiving location is far from the transmitter and the received signals are weak, a more involved structure may be necessary. In this case several sets of rods may be used to increase the sensitivity of the antenna. In cities, signal reflections from near-by buildings may be very troublesome in producing ghost images, and in this case also a complex antenna structure may be necessary, in order to discriminate against the reflected waves. In any event the installation of a television receiving aerial is much more of a problem than that of a conventional radio aerial. It is usually wise to have the installation made by a competent service man.

What About Improvements in the Future?

Many prospective purchasers of television receivers have hesitated to make the plunge because they fear that the equipment may very quickly go out of date, if not become completely obsolete. This fear has been fed by widely publicized statements to the effect that any change in the manner of transmission at the broadcast station will render all receivers useless, that is, until the receivers are modified to accommodate the change. This is not strictly true, because very considerable improvements have been made in transmission since the inauguration of public television service, and all receivers have benefited. But it is true that if the number of lines in the scanning pattern, or the rate of repeating the pic-

The program director and the control engineers govern their decisions of what is transmitted by the content and quality of images appearing on monitoring kinescopes in the control room. (COURTESY NBC)

A giant miniature set built by William C. Eddy for a television program. When televised this appears to be full-sized. (COURTESY NBC)

Television embraces all the dramatic arts—a scene from the television production of the Broadway play "When We Are Married"; Mordkin Ballet class demonstrating dance technique; opera made its television debut with *I Pagliacci* March 10, 1940; the television camera follows the antics of Tom Howard and George Shelton during a variety program.

tures, is changed, most commercial receivers are incapable of following the changes and major modifications must be made in the receiver circuits. Moreover, the amount of pictorial detail which a receiver can accommodate is definitely fixed, and if the broadcast station increases the amount of detail beyond this limit, no improvement is noted in the received pictures.

These facts have made necessary the establishment of standards of transmission which will serve as the basis of all television broadcasting for a long period of time. The standards first used in this country were those drawn up by the Television Committee of the Radio Manufacturers Association. These standards guided the experimental operation of the television stations and formed a basis for the recommendation of the Federal Communications Commission that the industry continue research on that important question. Accordingly the engineers of the radio industry, organized as the National Television Systems Committee, made an exhaustive study of the question of the standards for television broadcasting and submitted their recommendations to the Federal Communications Commission. The F.C.C. adopted a set of standards based on these recommendations and authorized commercial television programs to begin on July 1, 1941.

These new standards fix the line and frame frequencies at 525 and 30, respectively. The 525 line image provides for greater detail in the pictures transmitted than the previous 441 line standard. This change will be helpful in view of the trend to larger screens on current receivers. The use of frequency modulation is required for the sound channel accompanying the pictures.

Thus television is benefited by the recent developments of frequency modulation. These standards, observes the Federal Communications Commission, "represent, with but few exceptions, the undivided engineering opinion of the industry. They satisfy the requirement for advancing television to a high level of efficiency within presently known developments."

The fact remains that standardization does have the effect of preventing large-scale improvements in the pictorial detail of the pictures. In the future there will no doubt arise a demand for more detail, and the industry will then be faced with the problem of making progress without violating the standards. The most practical suggestion for overcoming this difficulty is the proposal to open up a new portion of the ether spectrum for a totally new service, while maintaining the old service on the old standards until public demand for it no longer exists. The F.C.C. has already laid the groundwork for such a wholesale expansion by assigning channels for television research which are not now useful for public program service, and by urging research organizations to develop their new ideas on these as-yet-unused channels. By some such scheme the problem of standards vs. progress can be circumvented without loss to the public. In the meantime the purchasers of television receivers can be reasonably certain that their equipment will not be soon out of date or completely obsolete.

From the preceding discussion it may appear that television consists mostly of problems with partial solutions. This is to some extent true; certainly many compromises have been made. But sight must not be lost of the central facts in the case: that a program can be sent which has the quality of a 16-mm. home movie (and a sound accompaniment considerably superior to that of most professional movies); that the television camera can pick up any subject on which it can be focused, if the subject can be photographed by the motion picture camera; that ultimately about 55 per cent of the nation's population (living in or near cities) may have such a service from at least one station, and a smaller percentage may have a choice of programs; and that all this is possible under conditions of transmission which were considered impossible even so short a time as ten years ago. The progress made since that time is, of course, a good indication of the progress to be expected in the next decade. In the meantime, the system is available and is being expanded both geographically and economically so that more and more people may enjoy its unique power to bring the outside world directly into the home.

CHAPTER THREE

The Men Behind the Camera

BY O. B. HANSON
Vice-President and Chief Engineer,
National Broadcasting Company

THE title of this chapter designates, in a general sort
of way, many men and many things. It is, in short,
a way of describing the individuals engaged in technical
production and the instruments with which they work.

Where the artistic interpretation of the director ends
and the technical interpretation of the engineer begins
is a very nice question. From one point of view, the entire problem of production rests squarely on the shoulders of the program director. Certainly he is responsible
for the artistic presentation of program material before
the television camera. Yet, it is manifestly impossible for
the director to produce his television program alone.
Without the active assistance of the entire technical staff
many important details would go unattended.

Let us, however, attempt to mark the line of difference
by defining the responsibility of the technical staff. The
television engineer takes responsibility for televising a

scene and transmitting it with maximum fidelity. With adroit camera work, good lighting technique, and careful attention to the important activities of the control room, the engineer contributes much to the ultimate success of the best television material. But he cannot make a good program of bad material. It is, therefore, the duty of the technical staff to advise, and to place at the disposal of the program director the best in television technique. And to be of greatest assistance, the engineer must not only master his own craft but also possess much intelligence that is of a purely artistic nature.

The observant man or woman quickly becomes aware that all media of entertainment, whether the legitimate stage, motion pictures, or radio, have their limitations. That we ordinarily do not become aware of them is a tribute to the ingenuity and skill of both technical and artistic directors in working within those limitations. The legitimate theater, for instance, frequently finds itself cramped for want of space on the stage and the height of the proscenium. For this reason directors have long since devised a group of traditional artifices. The actor may, for instance, look out of a window to witness something that is supposedly taking place in the street below. A newly arrived character rushes onstage with tidings of an event just completed at some remote point. Radio is sometimes impressed to deliver a description of some happening indispensable to the progress of the plot.

Even onstage, space frequently becomes too skimpy for the accommodation of a large cast or an even larger set, so that both must be squeezed within the available area. Conversely the stage, being of very definite proportions, often presents the strange sight of a small bedroom about thirty feet long and twenty feet deep.

Motion pictures, although they have enormously pushed out the limitations of theatrical art, are nevertheless themselves limited. The fineness of the film's grain sets a definite boundary to the amount of detail that can be crowded into a frame, the intensity of a light beam in the projector likewise sets bounds to the detail that can be thrown onto a screen, and the character of the screen determines the detail to be reflected to the eyes of the audience. The close-up, first used extensively by the great David Wark Griffith, has revolutionized film technique. In itself, however, it was an attempt to overcome the limitation of detail possible in the long shot. Even today close-ups of two participants in a conversation are seldom used. On the contrary, we usually find a close-up first of one character, then of the other. And sometimes the emotions reflected on the face of a listener are accompanied by the out-of-frame voice of the speaker.

The Men Behind the Camera

Here is another limitation turned to artistic use—suggestion.

Television inherits the rich theatrical tradition of a thousand years, and all of the ingenious technique developed by motion pictures during the last forty years. I do not mean to say that we are able to make use of all of it. The only way television can reproduce, let us say, the battle scenes from *The Birth of a Nation,* is from a film sequence. And the scope of some of the larger scenes on the stage is still beyond the limits of our television system. The qualities of the iconoscope and the kinescope set definite bounds to the television production, just as do the various instruments used by motion pictures. Television studios are small and their lighting systems, meshed into the camera routine, are not infinitely flexible nor perfectly efficient. And beyond these lie a series of amplifiers, transmission lines, and, ultimately, the transmitter, the antenna array and screen size in the home receiver to stay the hand of the technical director at a measurable and irritating point.

One could easily, of course, point out the imminence of many techniques in television that are now a part of standard practice in motion pictures. The "process shot," as a single example, should soon be incorporated into television studio practice. Many other possibilities will remain unrealized for years to come. We shall not soon have the wealth of technical equipment and personnel that Hollywood has at its disposal. Our future studio plants and equipment must be the accretion of years.

Television is, like extreme youth everywhere and always, poor, inexperienced, and ambitious. Its ambition is laudable. And if television has had the misfortune to be started on its career at a time when money is none too plentiful, it must simply make the best of it; perhaps if unlimited funds were available its progress would be slightly faster. But experience is another matter; experience, more valuable than a fat purse, is to be got only the hard way. No amount of money will help the director solve the puzzle of a certain camera angle that surpasses the technical limits of the best iconoscope; neither will it help the field crew size up a situation at Madison Square Garden on a fight night. Ten years hence we shall all chuckle over today's technical routines, just as any man finds amusement in the ungainly antics of his youth; but today, when we are climbing painfully over successive obstacles, we find our behavior anything but amusing.

Since it is with today that we deal here, I shall give the reader a peep behind the scenes of technical production as it exists in the National Broadcasting Company studios at Radio City. We shall have to do with studios, lights, cameras, titles, control room activities, and the actual mechanics of television broadcasting. And outside the studio walls, the subject of mobile unit pickups and their peculiar conditions will come up for brief consideration. Studio production of a dramatic presentation, television's equivalent of the full-length Broadway show, will perhaps illustrate more of the problems than any other type of studio production.

Television's technical staff is a tightly organized unit that includes cameramen, microphone boom operator, lighting engineer, video and audio control engineer, technical director, and various individuals concerned solely with the transmission of image and sound signals. Close organization is imperative; when the staff goes into action any mistake, any failure of co-ordination, instantaneously reflects itself in the technical quality, and hence the artistic appeal, of the television presentation. Let it never be forgotten that there can be no retakes in television; the first chance is also the last.

Technical considerations enter television production from the very start of the program director's speculation on the drama, variety show, or education feature he should produce. For the studio is of finite dimensions; no earthly power can expand it a single inch to accommodate any drama. And just as the studio in which the drama is to be televised has definite proportions, so there are but three cameras at present with limited space in which to execute the camera routine. A certain amount of illumination may be disposed about the studio from a power supply which measures in the neighborhood of fifty kilowatts. All of these factors enter, or rather should enter, into the selection of the script.

Having selected his script, however, the program director calls upon the help of the director of technical production to go over the technical needs of the forthcoming production and either to confirm his opinions of technical possibility or to learn the worst. The lighting engineer makes his appearance for a consultation on whatever dynamic effects the director has in mind. And special effects, such as snowstorms, gun shots, fire, wind, and so on, come up for discussion and disposal. A scheme for titling the production is outlined. Before the series of conferences is ended a technical director, who also acts as camera supervisor when the drama finally takes the air, has been assigned to work hand in hand with the program director.

With the program director, when he calls his first rehearsal, goes the production's technical director, always with script ready at hand. I might add that cameras are nowhere in evidence during most of the rehearsals. These, in fact, proceed in a rehearsal hall, following a practice which is rather common in the legitimate theater. Props consist of tables and chairs, usually, and the sets are merely slats on the floor marking walls, doors, windows, and so on. Cameras, of course, are purely imaginary at this point.

For ten days such rehearsals continue, the director altering his script and, with the aid of the technical assistant, planning the camera routine. Always keeping in mind the relative positions of his cameras during the rehearsals, the technical director advises which shots are possible and which are beyond the limits and abilities of men and technical equipment.

These are obvious products of rehearsal. More important, probably, is the fact that during these days of rehearsal—and they are fantastically few by comparison with legitimate theater rehearsal days—the director imparts the spirit of his production to the technical director. So much so that, once the drama goes into the studio for its final rehearsals with cameras, the technical director has not only mapped out the general plan of camera attack, but also visualized the scenes, mastered their relationship to each other and interpreted the drama to his own satisfaction and, we hope, the director's as well, in terms of cameras, lights, and movement. This last is peculiarly the province of the technical director. The program director thinks primarily in terms of artistic production—groups of actors, lines, stage business, costumes, sets, and props. The technical director is the representative of the television system, as such, sitting in on the rehearsal. His main concern is that the production shall squeeze within the limits of the system, and that men and matter shall faithfully transcribe the production as it shapes itself before the cameras.

So far only a single representative of the technical staff has become intimately acquainted with the production. The others—camera and microphone operators, lighting technicians, sound and image control men—either do not know the play or the director's conception of it, or their contact with it has been of the sketchiest. The reason for this condition is twofold: First, television is at present operating with a minimum of staff and, second, it has at present only a single studio from which live talent productions are put on the air. Our drama, therefore, really takes shape with the assistance of no more than one member of the technical staff.

The piece begins, in the lay mind, to resemble a show only when it comes into this studio for the final two days of rehearsal. The set, or sets if there are several, have been assembled and dressed and cameras and cameramen are ready for work. Overhead, a hundred lamps slant their rays down on the scenes, and other portable units are maneuvered to give depth to the image. The boom microphone stands awaiting the dialogue.

Studio cameras used at NBC are of the iconoscope type. Three are available for each studio production.

The Men Behind the Camera

They are enough like the conventional camera for no extended description to be necessary. Each has a double lens system: one for focusing the image on the iconoscope plate, the other affording the cameraman a view of the scene to be televised on a ground glass. Two of the cameras are mounted on mobile pedestals, with silent motors to raise or lower the cameras, which, in turn, are fixed to "panning heads" of the motion picture type. These cameras are used only for stationary shots, moving from position to position according to the shooting script.

Our third camera, identical with the other two, is mounted on a standard motion picture camera "dolly." It requires, therefore, the services of a man to move the dolly. But, being heavy, the dolly camera may be used for moving shots, such as, for instance, a sequence where the viewer is apparently carried from a distant point into a close-up of the action, or the reverse wherein the close-up gradually becomes a medium shot. This practice is a familiar one in the movies.

All of these cameras, be it noted, may be raised, lowered, tilted through a wide angle, and "panned" from side to side. They may be used with a variety of lenses, from six and one-half inches in focal length, to an eighteen-inch lens assembly. Ordinarily, one camera carries a lens of the shortest focal length for "long" shots, while the other two are supplied with longer focal length lenses for close-ups. Owing to several factors, the attainable depth of focus is not great, making it necessary to confine stage action within rather restricted areas.

These are the cameras. Their function lies in converting the light image focused on the iconoscope plate into a train of varying electrical impulses. These impulses are carried off to the studio control room over a coaxial cable.

It may be well here to say a few words about camera technique. Television has inherited what Hollywood learned most painfully over a period of more than forty years. The story technique introduced by Sidney Porter in *The Great Train Robbery* of 1903, the first complete story to be filmed, naturally was taken over by television. The "editing" feature, even older, also belongs to television. And the close-up, by means of which Griffith definitely severed the movies from stage technique in the days before the first World War, has been borrowed by television. Television, therefore, begins with an invaluable technique contributed by motion pictures.

But there are differences, the chief of which lies in television's kinship to the legitimate theater. Its scenes are sustained, building up to a climax as they do on the stage. Hollywood production, on the other hand, is piecemeal. Sets are built and lighted and shots are made over a period of days, frequently weeks. In television the same result must be achieved in minutes since camera routine must fit the sustained character of the acting. In television production we have the timing of the stage with the complications of photographic technique thrown in.

This is a very important point; it explains why three cameras are used simultaneously. All operate continuously, but the output of only one camera at a time is put on the air. In this way television finds itself able to present the variety of close-ups and long shots and angles that the movies have made desirable. And it also explains how television switches instantaneously from one scene to another.

For convenience, the several sets against which the piece will be played are arranged in logical order about the studio walls. The three cameras are deployed, usually in a roughly semicircular order, in front, let us say, of the first set. As the drama unfolds, the output of first one camera then another of the battery goes on the air. Then comes the time for the switch to a second set. One of the cameras is detached from the group and moved to the next set. At the conclusion of the last line on the first set, either a fade-in or a direct cut is made to the camera now focused on the second set. The other two cameras quickly move into position and soon the entire battery is working before the second scene.

It is difficult to state in so many words just what are the duties of the cameraman, and it is particularly difficult to convey to the reader the importance that attaches to his job. He works largely at the direction of the program and technical directors, seated in the control room adjoining the main studio. The cameraman is called upon to frame the scene with due regard for pictorial composition and to keep the action in focus. And, at the conclusion of his shot or shots from a given position, he is expected to move his camera to his next position, as indicated on a cue sheet affixed conveniently to the end of the camera. A green light tells him when his camera is on the air; once the light goes out he is free to move his instrument.

Stated in this way, the cameraman's task appears to be simply one of following orders, an almost purely mechanical job. Nothing, of course, could be further from the truth. Actors, being human beings, may never perform twice running in exactly the same way. Realization that their performance is at last going on the air causes them to make slight improvisations, to move a step further in this direction or that, and to change their stage business in a dozen ways. Usually this is all to the good as far as the viewer is concerned; it makes for a more interesting performance. But, from the cameraman's viewpoint, each change from the routine of the dress rehearsal constitutes a challenge to his ingenuity. I can best illustrate by citing two instances, in one of which the camera routine

The director instructs his cast during a break in the day's rehearsal. (COURTESY NBC)

The stage manager's job is the important one of carrying out the director's instructions relayed to him from the control room by a telephone connection. (COURTESY NBC)

Earle Larimore and Marjorie Clarke in *The Unexpected,* the drama presented during NBC's "First Night" program, May 3, 1939. (COURTESY NBC)

An unretouched image photograph of a scene from *Little Women,* made by O. B. Hanson at his home in Westport, Connecticut, forty-six air miles from the transmitting station.

broke down entirely. In a variety television broadcast, the announcer, apparently feeling that he was free for a moment, wandered off the set. The act in progress ran through more or less on time and very much as it had been rehearsed, but at the conclusion—no announcer. The camera assigned to the job of picking up the announcer began searching wildly about the studio, giving the viewer the effect of being in one of the "crazy" houses that used to be a popular feature at Coney Island, in which walls, floor, and ceiling began tumbling about. Fortunately, the cameraman knew where the announcer would have to reappear and he finally focused on that point, picking him up as soon as he made his reappearance in the studio.

On another occasion an arc lamp which was placed in position to give an artistic effect suddenly failed. Over the private telephone line came the message from the technical director, "Listen carefully, the arc is gone and we're going to have to do this whole scene without it." And the whole pickup was done differently from the rehearsed routine, without, I daresay, a single viewer being aware that a change had been made. These two instances illustrate the task that confronts the cameraman better than any formal statement I could make.

In addition to cameras, television needs studio lighting. The system chosen must not only provide sufficient illumination for the technical needs of the camera; it must also be adequate to meet the demands of any artistic effect desired by the program director. This latter becomes increasingly important, with the development of iconoscopes of greater sensitivity. Finally, unlike the lighting systems used in Hollywood production, the lighting system must be fluid in action. The motion picture director and head cameraman set their light units for the shooting of one particular sequence; once that is accomplished, production is stopped and lights are rearranged for the following sequence. In television the shift of lighting emphasis from one scene to another must flow as easily and smoothly as the camera action itself.

To meet the needs of such a system, NBC experimented for years with various types of lamps, gradually progressing through all the types that serve the needs of the movies, until our lighting engineer developed a system that meets both the requirements of television as to mobility, and the air-conditioning system installed in the studio. The system consists essentially of ceiling units of six special incandescent lamps, usually of five hundred watts each, which may be swung through a considerable angle. Each individual unit of this key lighting system is remotely controlled from a light bridge, making it possible to focus any combination of units on any particular spot in the studio. The system is completed by floor units of similar design manned by lighting technicians.

The lighting system, together with the battery of three television cameras simultaneously exposed to the scene, I believe is unique to television. The system of illumination, as I have pointed out, had to be devised to meet the needs of the curious hybrid of production—legitimate stage continuity of action with motion picture type of camera routine. This latter, in turn, had to be devised to supply the viewer with the interesting variety of angles and shots that he is accustomed to find in the movies.

The lighting system and the cameras are the main items in the technical equipment of the studio proper. There are, in addition, any special effects, such as snowstorms, wind machines, fire effects, and so on, demanded by the program director, and one of several mechanical methods of titling the production.

The nerve center of the television production, however, is outside the bounds of the studio proper. In the control room, directly adjoining the studio and located at an elevation of about ten feet above the studio floor, sit four men who direct all activities on the set. Here are the program director and his technical director, seated side by side at a desk commanding a view of the monitoring kinescopes in the wall directly in front of them.

Here also are the sound control and image control engineers, the one corresponding to Hollywood's "sound mixer," the other maintaining close watch over the contrast range and over-all illumination in the pictorial output of each camera.

The unifying link between cameraman and technical director, program director and stage manager, and sound control technicians and microphone boom operator, is the private telephone line. By means of such a circuit the sound control technician governs the action of the microphone boom operator, whose instrument hangs from an extension arm hovering just above the range of the camera. So too does the technical director communicate with his three cameramen on the floor of the studio. The program director, finally, can communicate with his stage manager on the floor to wigwag the actors to one side or another and prepare them for the cues from the beginning, when the production goes on the air, until the final flourish indicating that the show is ended.

The control room is, as those who are familiar with motion picture technique will have perceived, the equivalent of the Hollywood cutting room. But with quite a difference! The editing of the television production is performed while the production is on the air. In Hollywood editing it is accomplished after actual shooting and laboratory work have ended; it is one phase, separated entirely by space and time from other phases. In television, "control" is the central point of production which includes every phase—camera work, lighting, titling, editing, and distribution. Time is, therefore, of the essence. And split-second decision, governing all the co-ordinate parts of production, is the hard rule by which television men live.

The one man on whom this burden bears most heavily is the same technical director who has accompanied the program director from the first rehearsal. Seated at the elbow of the program director in the control room, it is he who relays all directions from the program director to the cameramen, adds a few of his own, and relays the total to the cameramen. He makes the switches, the output of first one camera, then another, into the outgoing channel by means of a series of interlocked push buttons.

The director and his technical chief must, of course, have some sort of guide to govern their decisions. The actual appearance of the studio and its cameras cannot be of much help, since the human eye embraces a far greater area than does any one of the cameras. The control room staff, therefore, is provided with a series of monitoring kinescopes, which reproduce the images actually being registered by the respective cameras. Two of these are at the disposal of the program director and his technical aide. The first shows the image actually in the outgoing channel, that is, the image being broadcast from the transmitter in the Empire State Building tower. The second monitoring kinescope, immediately to the right of the first, is a preview instrument reproducing, at the will of the director, the image being registered by either of the two remaining cameras on the floor. In this way the director may call for changes in composition, focus, and angle up to the instant the camera is switched into the outgoing channel. The third monitoring device is placed at the disposal of the video control engineer that he may make shading adjustments in any of the three images being picked up in the studio.

There is one studio, the film scanning room, of which I have thus far omitted mention. It is, of course, a highly important element in television programming and one that deserves more space than can be allotted to it here. In it we are able to scan film of both the standard thirty-five-millimeter and sixteen-millimeter sizes. And also we are able to insert film sequences into the live talent productions in the studio described above so smoothly that one is unaware of the transition from one medium to another.

Beyond these rooms, and linked to them by coaxial cable, lies the equipment room, filled with amplifiers and containing the vital synchronizing generators. Another cable circuit carries the sight and sound signals from

The Men Behind the Camera

Radio City to the NBC transmitter in the Empire State Building tower for radiation over the entire service area, extending about sixty miles in all directions from the antenna array on the very top of the building, nearly a quarter of a mile above Fifth Avenue.

I have treated of cameras, lights, rehearsals, and controls at such length that the reader may have lost sight of the forest for the individual trees. I cannot emphasize too strongly that in television it is the totality that counts. Every individual must act in perfect co-ordination with his fellows. Inept camera work will surely ruin a production as quickly as inexpert handling of the controls or a failure of equipment. Perfect co-ordination and timing are the ideal, but even satisfactory television must not fall far below the ideal. With that in mind let us turn back to the studio and look in just before our drama goes on the air.

Fifteen minutes and more before the drama is scheduled to go on the air, all those involved in its presentation—actors, stage manager, cameramen, lighting crew, microphone boom operator, and the entire control room staff—have taken up their positions. Upstairs, in the film scanning studio the projectionist and the motion picture control technician have made final tests of their equipment. Cameras in the studio below have been checked, lights are in position for the first shot, and titles are in readiness for the opening. Minutes go by. The director points out to his stage manager that an actor is out of place. "Three minutes . . . two minutes . . . one minute." The technical director, now camera supervisor, relays the same information to cameramen and the motion picture studio complement, who are holding the opening leader, "RCA—NBC Television Presents," ready. Thirty seconds, then ten, and finally the hand jumps to the hour.

"Roll 'em!" the director shouts, and his words are echoed by his technical aide into the line linking him to the motion picture control desk.

Instantly the first frames of the film leader appear on the monitor of the outgoing channel. Seconds later the end is reached.

"Fade out film—fade in two!" Camera number two's green lamp is lighted and its image replaces the film, and whatever title device is used appears on the screen. While the titles succeed one another the first shot of the actual play is previewed on the second monitor, and whatever corrections seem desirable are made. The title ends.

"Take one!" And the camera supervisor switches the "dolly" camera on the air. The stage manager points to the waiting actors, and the dialogue begins.

"Take two! Tell three to pan right a little." Instantly camera number two is cut into the outgoing channel and the direction is relayed to the cameraman on number three.

Directions follow one another, communications fly over the private telephone lines, background music is dubbed into the production from a turntable in the control room. Sound is monitored and the correct effect is introduced; it would obviously be ludicrous to have close-up sound in a long shot.

With one thing and another, the control room is filled with clashing sound—dialogue, music, and directions mixed—so that the ordinary mortal finds it difficult to believe that a television production is going on the air. Only a tenseness in the atmosphere betrays the taut nerves that are the price all must pay who engage in television production. I doubt that it will ever be otherwise. Taken altogether, television production was correctly appraised by a man thoroughly conversant with the legitimate theater and motion pictures. "It's the strangest and most amazing form of theater I ever saw," he said.

Shot succeeds shot, the green light leaps from one camera to another, long shots become close-ups, sight effects and sound effects blend into the production, light units on floor and ceiling swing about to follow stage action, until, finally, with a last arm flourish the stage manager signals that the studio is off the air. The show is ended, with only empty sets and worn scripts to testify that it had ever lived either in the studio or in the homes of thousands of televiewers.

There is, however, another highly important type of television broadcast. It is the outside telecast, made from a sports field, an indoor arena, a street corner—any place, in fact, where television cameras can penetrate to cover a significant or interesting event. NBC, in less than a year of television broadcasting, has covered the opening ceremonies of the New York World's Fair, major league baseball games, intercollegiate football and baseball contests, boxing and wrestling, track meets, parades, the activity at several airports, and other events too numerous for mention here. Practically each such program has constituted some kind of television "first." And a most spectacular "first" was the transmission of aerial views of New York City from a plane several thousand feet in the air.

The field pickup and relay are made with a mobile television station. The units of this station are two: a mobile equivalent of the studio control room, together with two camera chains; and an ultra-short-wave relay transmitter for use where no line connection links the point of pickup with the main transmitter in midtown Manhattan. An experimental wire circuit, for instance, is ordinarily used by NBC in relaying programs from Madison Square Garden to the transmitter. In making pickups at the Garden, therefore, it is unnecessary to use the relay transmitter.

The Men Behind the Camera

From Ebbets Field, in Brooklyn, on the other hand, no such line circuit exists, as the use of telephone circuits for transmitting vision signals is at present possible only over extremely short distances, and the relay unit must accompany the mobile control room. The circuit from the point of pickup to the main transmitter is a radio beam, of a wave length sufficiently far removed from that of the main transmitter that it does not interfere with televiewers' home reception.

The problems encountered by the field group are obviously far more numerous than those inherent in studio production. The site of the pickup must be surveyed for power supply, camera positions, an antenna location, light, and transmission path, not to mention the highly important factor of parking space for the two large motor vans in which the equipment is mounted.

The tussle of the field group is with time, space, the elements, and unrehearsed program material, a situation not unlike that usually confronting the newsreel camera-man. The difference, of course, lies in the fact that the entire product of the field group goes on the air; the newsreel is edited before it is shown to the public. And very frequently the crew, which numbers an engineering staff of cight and a program director with his assistant, labors under the most adverse conditions. In Washington, D. C., early in 1939, they transmitted images during heavy snowstorms. They covered the parade of the Army's "Iron Horses" down Fifth Avenue in a downpour. And at Madison Square Garden recently a satisfactory pickup of a National Hockey League contest was made under light so feeble that a newsreel cameraman would have found it difficult to work.

The technique of the outside television broadcast differs, as I have said, from studio practice. With camera positions chosen, lines are laid to connect the instruments with the mobile control room. One camera is usually fitted with a close-up lens; another usually covers the entire field, or as much as can be embraced within the limits of the camera. At Ebbets Field, in televising the double-header between the Brooklyn Dodgers and the Cincinnati Reds, one camera scanned the home plate area from a box behind the third base line. The second camera had its lens trained on the field from a second tier box.

Private line telephone circuits again connect cameramen and announcer with the program director and field supervisor inside the mobile control room. This unit, together with the transmitter van, in the case of the Ebbets Field pickup, stood in a street outside the park. A directive antenna array sent the electrified program over a radio beam to the main transmitter, where it was demodulated and impressed on the radio carrier of Station

W2XBS before being broadcast from the antenna on the top of the Empire State Building.

These two large motor vans are obviously not suited to all types of television relay. In many instances we have found it impossible to pick up interesting program material because of the bulkiness of equipment. Accordingly, we set about remedying this condition in 1939, with the design of "vest-pocket" field equipment. This apparatus, described fully in the chapter, "The New Newsreel," may be transported in a rather small vehicle to any field point. Or it may be used in picking up some of the interesting radio programs being sent out on sound broadcast channels from Radio City. We expect this equipment, the smallest television unit yet built, to enlarge the scope of field pickups to a very considerable extent. This marks, I believe, a very important step toward the ultimate goal of carrying television's electronic eyes to every site of an important or interesting event.

I hope that I have not conveyed the impression that I consider television a fully, or nearly fully, developed medium. Future developments will make our present achievements seem very crude indeed. New pickup tubes, of which the orthicon is a forerunner, will vastly increase the sensitivity of the television camera. The "vest-pocket" equipment of which we are so proud today will probably seem elephantine to us ten years hence. There will be mobile units with self-contained power supplies to free television from dependence on power networks. One can anticipate field cameras of almost infinite flexibility. All of these things, together with great studio plants and nation-wide networks, are certainties. And with them will come new production techniques that will make today's practices seem like those of Edison, Biograph, and Lubin in the old movie days.

CHAPTER FOUR

Programming

BY THOMAS H. HUTCHINSON
Program Manager, Television Department, NBC

THOSE who have witnessed television programs over our New York City station must have arrived, however circuitously, at the conclusion that some method underlay their selection. For, despite the apparently distant relationship between "skin acts," jugglers, harmony teams and dramas, football and basketball and baseball games, at one time or another the televiewer becomes aware that behind the heterogeneous agglomeration of entertainment lies some sort of plan. Usually, for instance, the televiewer can be reasonably certain of witnessing a play on Friday evening, and usually he finds himself confronted with variety of some sort on Wednesday evening. And Thursday afternoon is pretty certain to bring a motion picture feature.

Now, the unfortunate individual who stands behind this plan of television entertainment, the man responsible for the selection and arrangement of the week's schedule, is the manager of the television program division. The individual of whom I speak is at once everything and nothing at all. On pleasant days when rehearsals are proceeding on schedule and next week's program list is complete, and last night's transmission was, on the whole, very satisfactory, he is the Sam Goldwyn or Jesse Lasky of television, lord of all he surveys—studios, directors, actors, and all. On most days, however, when each telephone bell, every letter and caller is a warning of harsh criticism and complaint, he is "it" in a very exacting and enervating game.

The root of all the program manager's despair is the simple fact that he has a certain amount of money to spend, and certain responsibilities that attach to the spending. The money he has wheedled, with the loyal assistance of his chief, from the financial guides of the company which has seen fit to employ him. And of course the sum is never large enough. Nevertheless, every program director comes laden with plausible arguments why the share allotted to the production he has in hand should be larger than it is. The man in charge of the selection of motion pictures hears an exciting tale of a wonderful feature that is virtually a "steal," except it costs a little more than the last movie feature. And booking agents, who also have heard of television, clamor for attention. The respective "acts" they represent carry excellent press notices and should "go big" in television. In fact, they are just what television needs.

It is the manager's duty to say a loud No! to practically all of these propositions. Not, in many cases, because the agent's or director's ambitious plans have no merit. The act may be all the agent claims for it and the director may be fully competent to use every cent of the additional budget to good ends. There are, however, thirteen or fourteen other hours of television broadcasting to be filled with as interesting material as it is possible to procure. And all must be done within the limits of a very limited budget.

Between conferences bearing on such matters, the program manager must find time to arrange rehearsal schedules, clear legal rights to dramatic productions, approve or disapprove proposed scenic designs, and generate a few program ideas on his own.

The basic problem that besets the program manager is stated quite simply: What is a television program? Therein lie both hope and despair. No one knows for sure what makes a good television program. There are as yet few rules for the building of either a television program or a program schedule. We may hazard a few guesses, basing them on the accumulated experience of a brief period of television broadcasting. We may measure certain trends from the reactions of our present small television audience. But even these trends may change with the sale of many more receiving sets, and the consequent growth of an audience reflecting a wider variety of tastes.

We work in the hope that we will one day arrive at just that happy blending of the old with the new which will make television programs a special field of entertainment and education. This requires both boundless patience and the courage to face many a bitter disappointment. On occasion, after weeks of careful preparation, we have had to admit that a certain play was a failure as a telecast. It may have appeared to be headed for success during rehearsal. It may have been acted smoothly; the camera treatment may have been skillful. Yet, the total of all this was not good television entertainment.

On the other hand, we frequently have surprising successes, in presenting some little-known singer or monologist. At times, this has been due to some harmony of personality between artist and viewer. Hildegarde is an almost perfect example of this sort of thing, although she had international fame long before she stepped before NBC's television cameras. Gilbert and Sullivan operettas are another source of almost sure-fire television appeal. Plays, again, are another matter, although we all felt reasonably certain of the success of such a drama as *Missouri Legend*. But, surprisingly enough, some educational features, such as Dr. Georg Roemmert showing and commenting on the microscopic life in a drop of water, and Betty Crocker making an apple strudel, were also television hits.

All of which brings us back to the basic question, What makes a good television program? Quite frankly, we do not know all the answers, but we do know some of the basic essentials. Television can tell a story, with a technique which is a hybrid of stage, motion picture, and radio practices. It can also explain a process or portray the results of a process, which is the basis for many an interesting educational feature. Television is also capable of simple transmission of entertainment previously prepared for other media, such as vaudeville and motion pictures. Finally, its mobile units can rove about to relay actualities occurring in the city streets, in an indoor arena or an outside field. Studio shows, motion picture transmissions, and outside programs have, therefore, been the basis of our program experimentation since 1936.

Since 1936, and up to the present, we experimented with all manner of talent—drama and dancers, musicians and magicians, comedians and acrobats, and what not. We have found that practically anything the legitimate stage has developed can usually be televised successfully. We found also that certain types of artists and entertainment lent themselves to the new medium more readily than others. We felt, however, and we still feel, that all types of entertainment should be tested before the cameras to ascertain their television possibilities.

Our first programs presented only one or two artists before the camera. From such a simple beginning we have progressed in a little more than three years to elaborate productions with a running time of more than two hours. Casts have included as many as sixty performers. Musical comedies, dramas, fashion shows, and variety give our studio programs a broad scope. Outside the studio, the mobile unit group has steadily pushed back the horizon of possibilities. The recent transmission from an airplane demonstrated in a spectacular way what television of the future holds in store for its audience.

Let us examine a single week's program schedule as transmitted over NBC's Station W2XBS:

| Sunday | 3:45- 5:30 | Amateur Hockey: New York Rovers vs. Valley Field of Quebec, at Madison Square Garden. Bill Allen, announcer |
| | 8:30-10:30 | "When We Are Married," a telecast of Robert Henderson's production of J. B. Priestley's comedy, currently at the Lyceum Theater, with Estelle Winwood, Alison Skipworth, J. C. Nugent, |

		Tom Powers, Ann Andrews, Sally O'Neil, and A. P. Kaye
Wednesday	3:30	"Miracles of Modernization," an FHA film
	3:40	"Yours Truly—Ed Graham," film
	4:00- 4:30	Coward Shoe Style Show
	6:45- 7:00	Lowell Thomas, news commentator
	9:00-11:00	Boxing: Golden Gloves Tournament of Champions, at Madison Square Garden, Sam Taub, announcer
Thursday	3:30- 4:30	"The Phantom Fiend," thriller film, with Ivor Novello
	6:45- 7:00	Lowell Thomas, news commentator
	8:45-10:30	Hockey: Boston Bruins vs. New York Americans, at Madison Square Garden. Bill Allen, announcer
Friday	3:30	"Washington—Shrine of Patriotism," film travelogue
	3:50	"Cliff Friend," musical film
	4:00	"Ten Minutes in Sweden," film, with Eric Mann, narrator
	4:20- 4:30	"Westminster Kennel Club Show," a Newsreel Theater film record
	6:45- 7:00	Lowell Thomas, news commentator

	8:30- 9:30	"Dangerous Corner," a drama by J. B. Priestley, with Ruth Weston, Alexander Kirkland, Helen Craig, Barry Thompson, and Helen Brooks
Saturday	3:30	The Children's Matinee
	4:00	"Jasper National Park," film travelogue "Sunday in Mexico," film travelogue
	7:30- 8:00	"Art For Your Sake," with Dr. Bernard Myers, a radio-television feature
	9:30-10:30	Knights of Columbus Track Meet, at Madison Square Garden. Jack Fraser, announcer

This schedule, which represents quite an increase over the two-programs-a-week task of our first weeks of television, pretty nearly touches the limit of our plant capacity, to say nothing of the programming staff. This staff consists of seven studio program directors, a director of outside broadcasts and his assistant, a man charged with the procuring of motion picture films, several stage managers, a scenic designer with his assistants and other individuals—twenty-two persons, all told. And small though it be, this group represents the largest television program staff at present operating in the world. Its size was determined largely by intelligent guess, being based on the experience gained in three years of experimental television broadcasting before the beginning of a regular service in 1939. Subsequent events have confirmed the accuracy of the estimate.

All of the directors of studio programs came equipped with a background of legitimate theater experience, with the sound knowledge of stagecraft that is therein implied. A few have had, in addition, extensive radio experience, and others have done their turn in the movie industry of Hollywood. They are men of varied talents and temperaments, some excelling in dramatic productions, others in musical shows, and still others in variety programs. Each has a different and quite definite temperament, which reveals itself in every television program he produces. Thus far, however, we have not attempted to impose specialization upon our program directors; each in turn directs the production of a drama, which is followed by several variety shows of different types.

The television program manager's task is expressed very largely in his relations with the members of his staff. In briefest description, he is television's equivalent of the producer in a Hollywood film company. He apportions the budget, allots rehearsal time in the studios, approves program ideas, sees to it that legal obstacles to production are removed, makes whatever suggestions on production that he can, and, in general, supervises each production from the time when it consists only of an idea through its dress rehearsal. This is a very large order, and it is obvious that the managing individual must rely to a great extent on the judgment, ability, and good taste of the various program directors. It is the program director who makes either a success or a failure of each individual program; the manager's main concern is with the general level of excellence of the entire schedule.

It may be well to detail the story of the program manager's part in the production of a drama for television, since this is the most ambitious of our works and the one around which the remainder of the schedule is built. The program manager starts things with the assignment of a particular director to the job of producing a dramatic show for a given evening some weeks in advance. The director probably has a script in mind and he also probably has a fair idea of how he will go about producing the piece. He outlines his scheme and assures the manager that not only is it possible to produce the play within the technical limits of the studio plant, but also that the production will come well within the limits of the budget. The play goes into the manager's brief case for after-dinner reading and a subsequent guess on what production will cost.

It has been estimated that the average Hollywood feature film of about ninety minutes' running time costs about $300,000. This, of course, is an average figure; many of Hollywood's elaborate productions are budgeted at $1,000,000 and more. In comparison with the motion picture industry's average cost of producing a feature film, we in television have found it necessary, and possible, to deliver a comparable item of entertainment, of approximately the same length, for $1,000. This amount is less than one-third of 1 per cent of the cost of the average movie film. This very low cost will undoubtedly rise when television broadcasting develops beyond this, its first, stage of development. Our production costs have been kept low because we have received the co-operation of artists, authors, and composers in launching the new medium. Royalty fees and actors' salaries, therefore, are far smaller than they will be in the future.

Nevertheless, the cost of a television program must always be a fraction of that of a movie feature. The television producer's job will be to deliver, with little time and money at his disposal, a group of programs that are appealing to both eye and ear. We have proved conclusively that this is possible. The television budget of the future may have many times the amount of money we now have to spend on our most careful productions, but the problem of delivering many hours of programs daily to a nation of televiewers must always call for the greatest skill in producing dramas, variety shows, musicals, educational programs, and outside events quickly and at low cost.

Having read the uncut script and assured himself that production of the drama is possible, the program manager approves the director's project. From that time forward until the scenic designer makes his appearance, the program manager's contact with it consists largely of routine reports. The director takes over, cuts the piece to approximately the desired length, casts the play, and takes the script and players into the rehearsal studio on the appointed day.

The scenic designer comes forward with his rough sketches after the rehearsals for the drama have begun. The basic evidence he presents is a floor plan of the studio in which our dramas are produced, showing the relative positions and approximate sizes of the various sets involved. The sketches reveal, when accompanied by much wordage and waving of hands, what compromises director and designer have made with time, space, and monetary considerations.

Scenic design for television is in such a state of flux that it would be useless to lay down definite rules. In a few short months, for instance, we have progressed from the point where a lithographed bookcase was as convincing on the television screen as the genuine article to the point where real bookcases and real books are a necessity.

Programming

All of this has been due, of course, to the rapid increase in the sensitivity of television's pickup devices. Greater technical advances will necessitate greater fidelity of detail.

Since the television image is in black and white, we have found it most satisfactory to paint our scenery in whites, blacks, and intermediate tones of gray, always working within the dynamic range of the television screen, which is not the same as that of the human eye. The eye, of course, distinguishes a far greater variety of shades from black, through gray, to white, than the television system. We have also found it advisable to paint shadows on the walls of the set, rather than create shadows by the use of stage lighting. These considerations, as I have indicated, are those obtaining in the design of scenery for today's television; the art is certain to develop very rapidly with the introduction of instruments of greater sensitivity into the television system and the construction of new studios along more spacious lines.

Having done his part in the settling of questions of program content and scenic design, the program manager frequently finds himself drawn into discussions of "video effects." These are the amiable subterfuges employed in television, as in motion pictures, as substitutes for sets which, if built along natural lines, would be far too large for our studios. A miniature set, for example, may be of a mountain with its surmounting castle, or it may be a suburban bungalow in a newly developed real estate project. Obviously it would be both impossible and impracticable, even if it were possible, to build such sets and move them into the studio. A scale model, however, does the trick neatly.

One of our early experimental programs was built around a scene from *Tristan and Isolde*. We wished to set the locale of the love scene. We solved the problem by building a scale model of the stage set. Then tiny figures were placed in it in the same positions as the artists would have taken when the performance began. We opened the scene with a full camera shot of this miniature set. Then the camera moved in to a closer shot, "panning" as it did so, to focus main interest on the two little figures seated on the bench at the side of the set. When the point was reached in the camera movement where the models began to fill the screen, we switched to a second camera which registered the live actors in similar positions on a stage setting which was a full-size reproduction of a part of the set we had just shown.

This artful device is quite successful and has been used time and again in television production. In this way we have shown exteriors with miniature houses, cities, and villages. We have shown harbors with battleships in motion as well as moving railroad trains and automobiles. When a model is built in correct perspective, very convincing illusions can be created.

Once these major details of production have been passed under the program manager's review, his reminders that the production is under way consist chiefly of bills that appear on his desk for any one of a hundred items found necessary for the drama. These continue arriving until the final records of the production have been filed away.

All of this description has concerned a single entertainment item, usually of about seventy-five minutes' length. There are, however, perhaps fourteen more hours to be filled for the week's schedule. About 40 per cent of the entire schedule consists of live talent entertainment presented from our studios at Radio City; aside from the week's drama, the remainder of the studio schedule consists of variety programs, fashion shows, educational features, and, lately, programs presented simultaneously over both radio and television. Roughly 30 per cent of the schedule comprises events relayed from points outside the studio by the field group operating our mobile television units. A final 30 per cent consists of film. These averages were obtained from an analysis of about 600 programs transmitted during the eight months of 1939, our first year of regular telecasting under public service conditions.

Our dramatic presentations have achieved such success with pioneer televiewers that some critics have been quick to quote that "the play's the thing." Now, to a certain extent this is true, but I believe that the implications are wider than most persons realize. Variety can and should be, to my way of thinking, something of a play. What I mean is that a variety show for television should not be a miscellaneous collection of separate acts for presentation without regard to their relationship one with the others. The articulated variety hour, built about a unifying idea and held together by some theatrical device, should somehow come under the heading of "play." It is this sort of thing that we have tried out recently in a program called "Out of the Kaleidoscope," in which various related variety turns were united by the device of a projection kaleidoscope. The kaleidoscopic pattern halted at a given instant and out of it materialized the following bit of entertainment. More of these devices should be possible in the near future.

The method of building a variety hour for television differs considerably from that of the dramatic production. Usually the idea for several variety shows resides in the producer's mind before the program manager assigns him to the production of one or more of them. Certain entertainment or educational items may already have been scheduled for each such program. These items, or acts, have been gathered from the thousand and one which are presented to the program manager—either by the performers themselves or by their agents. With his own idea and an act or two, the director completes the list of entertainment units in the program, writes his unifying script, rehearses the show, and puts the entertainment on the air.

These variety hours, of which there are usually two in the week's schedule, together with the drama production, just about exhaust the possibilities of both program staff and studio facilities. For the remainder of the schedule from Radio City we must rely largely on film which, fortunately for us, requires only projection apparatus for presentation. On the other hand, we have not been able to tap the huge reservoirs of film material in Hollywood's vaults. Television, being a medium of entertainment largely visual in its appeal, has naturally aroused a certain amount of apprehension among the producers of motion picture films.

The problem of our relationship with the motion picture industry may be outlined in a few words. Hollywood has film, and television can find very good use for much of that film, even the older feature releases and short subjects. These films represent, however, an enormous investment in the aggregate, and it would be too much to expect that Hollywood should give this store of entertainment to television for nothing. Television is, however, both poor and young; it could not at the present time make payment for these films commensurate with their cost. Some day perhaps, but not now, television will be able to make offers that will pique Hollywood's interest.

Meanwhile we are forced to rely on the films held by independent distributors and on commercial films. Our payment for independently held features and film shorts is, considering the size of our present audience (about 10,000 persons), very handsome indeed. For the commercial films, produced for, and usually carrying the message of, some industry or particular business organization, we pay nothing. The fact that an advertising message is transmitted is considered as more than just compensation for the use of the film.

These films are broadcast on two different types of occasion. A film supplies an afternoon program when the live talent studio is occupied by a group in rehearsal; an evening film show fills a spot which otherwise might remain vacant, because of the limitations of the technical staff's work week.

In addition to the dramatic shows, variety hours, and film transmissions—the three general groups into which all studio productions fall—there is a separate and entirely distinct type of television program. This is the program, usually of some scheduled news event, which is relayed from some point remote from the television studios by the mobile unit. I shall not go into the theory of the outside television broadcast, since it is treated under a different heading. My concern is with the source of program ideas and the disposal of them.

Here, too, the director is a figure of commanding importance. It is he who must produce not only programs that are actually put on the air, but far more ideas for future programs, many of which will never undergo the scrutiny of the electronic camera. His source of ideas is the daily newspaper, the magazine, lists of coming events, sports calendars, and conversations with his friends, as well as a fertile brain. Many ideas are generated; most of them are immediately discarded as holding too little general interest or presenting too great difficulties, either technical or financial. The ones that remain are then the subject of numerous letters, telephone calls, and personal interviews. If they survive this ordeal a survey of the site is made by the television field supervisor, who heads the outside technical staff. In this way New York City and its immediately surrounding territory is gradually being plotted for relay possibilities.

Some points, of course, are already established on the

television map. Madison Square Garden, scene of numerous boxing and hockey matches, basketball games and track meets and other events, has thus become a regular point of call for the NBC cameras. For many weeks we also transmitted wrestling matches and boxing contests from the Ridgewood Grove Sporting Club. Yankee Stadium, the World's Fair grounds, and Ebbets Field, having been the scene of television relays, also need no further survey. A few hours of test in advance of the actual program transmission is all that is required on the second and subsequent program pickups.

These details have ordinarily been considered by the director of outside programs before he presents his final choices to the program manager. Again the age-old questions of cost come up, together with a discussion of the probable technical success of the relayed program and the interest inherent in the proposed television program. Then the proposed program is compared with others on the week's schedule, in order that there shall not be too pronounced a leaning toward, let us say, sports. If the outside event fits well into the scheme of things it is then scheduled, and if the event is one to take place in the open air, both director and program manager give silent prayer for sunny skies.

Here then is the week's schedule completed. It embodies as great a variety of program types as program manager and directors can give it. And if all goes well the schedule should lift the audience rating for the week a few hundredths of a point.

I have dealt here with present techniques, achievements, and problems. The problems we have encountered have been almost without number. The creation of new techniques for television production has necessarily been based on experience gained in motion pictures, the stage, and radio broadcasting. And this creation is a mighty labor, greater than that which brought forth the talking picture.

Television is a realistic medium which presents to the viewer a spectacle to be taken literally; for the eye is most assuredly literal. Radio, with its tricks of sound effects and background music, whisks its audience in a trice from New York to Singapore; the listeners supply their own scenery and even costumes for the actors. I have yet to find a viewer who could do the same for a television drama. The upshot of it all is that television, having destroyed the worlds of the listener's imagination, must supply new worlds. Of its ability to do so I have not the slightest doubt. *The Jazz Singer* came as a thunderbolt to cast confusion among the producers of motion pictures more than a decade ago. There was much talk of the possible entertainment value of the movie with sound, and harsh voices were raised in criticism on every hand. The issue has long since been settled; the talkie swiftly took precedence over the silent film. The average American will also, I feel certain, eagerly accept television when it is offered to him on terms that he can meet. It is up to the program director to see to it that entertainment and educational material be presented in as fascinating a form as possible to hasten the spread of a new and great art among the people.

It may seem that I have dwelt largely on promises of the future. We all look forward to the future of television with the utmost confidence and an almost total lack of restraint. The vision is of new wonders brought within the experience of the average man. Today's television, however, needs no apology. Any medium which arouses such enthusiastic comment from its audience needs no other justification. We have already presented members of the Metropolitan Opera in a condensed opera performance; we have given the televiewer numerous notable dramatic productions, including the first television broadcast of a current Broadway show. This was the recent transmission of *When We Are Married*. We have had major sporting events in many lines of sport. President Roosevelt was one of the first political figures to step into the television screen in the American home. We may be pardoned for a just pride in our achievements.

Programming

Television, now that a practicable means of network-ing has been developed, has been supplied with the final implement necessary for the creation of what will even-tually be a nation-wide service. Future programming will be vastly different from what it is today, yet I feel cer-tain that some of the techniques and practices we now observe will survive to become guides for future produc-ers. It has been suggested that television will follow the course of motion pictures, with elaborate sets, armies of actors, and corresponding budgets. I cannot agree. The emphasis in television programming, I believe, will con-tinue to be on persons rather than on things. The per-sonality of the actor and the personality of the director, as expressed in his production, must always be more important than even the most elaborate detail of the setting.

As for the craft of production, it should grow far be-yond its present bounds. Television's scripts will assume a distinctive form once a group of writers have the time and opportunity to study the qualities of the new art. With wider audiences a greater variety of material will come within the purview of the television cameras. And the director, finding himself in new and more spacious studios and supplied with some of the amazing devices that are in their formative stages in the laboratory, will produce programs that will be neither of the stage nor of the Hollywood lot, but of television.

The hand that ultimately shapes the production of a television program, however, will be neither that of the program manager nor that of the director. The televiewer, with his finger on the tuning knob, will forever be the court of final appeal. All of television must be shaped to his desires.

CHAPTER FIVE

The Director

BY THOMAS LYNE RILEY
Television Director,
National Broadcasting Company

A SMALL group of men stood at the Players Club bar one night and one of them said, "So you're a television director?" and the man who was addressed admitted it.

"Must be interesting."

"Yes, it is."

"Well, tell me, what's it like—theater, pictures, radio, or what?"

"It's something of all of them plus something new— television itself."

"Think it's going to amount to anything?"

"I think it's going to amount to a great deal. That's why I'm happy to be in it."

"How do you direct television?"

The man thought a moment and answered, "Oh, you just do, I suppose."

The discussion lasted into the night.

We Present Television

Yes, one "just does" direct television, but this chapter will attempt to convey some conception of what the director has learned about television from experience and how he goes about gaining that knowledge.

Other chapters in this book tell of the electrical and mechanical workings of television; of its place in science and public relations; of its future as a social and educational force; of its commercial possibilities. This chapter, however, intends to explain how television productions get on the air from NBC's New York studios and to state what the director did before and during their actual presentation.

What is a television director?

He is the person directly responsible for preparing, producing, and, in many instances, selecting television shows. It is his task to mold every facility of the medium to accomplish the most effective presentation. He should be capable of supplying television with imagination and knowledge, ability and experience. He must, of course, understand how to produce desired results in the most efficient manner.

What is the television director's future?

It would seem apparent that the answer will lie entirely with the kind of leadership which will take over when television becomes important financially. There should be vast expansion in all departments of this new medium which will cause a great demand for imagination and creative ability. Wise executives realize this. The successful director will grow with television.

With what does he have to work?

First and foremost is the miracle of television itself and all its physical properties: cameras, lights, scenery, effects, and the staff of engineers which has the responsibility of getting pictures and sound on the air. The director has set designers, property men, stage managers, musicians, actors, entertainers, speakers, and everyone concerned with the actual program content of his show directly under his authority. He is the connecting link between the world of make-believe, which is show business, and the world of reality, which is engineering.

At NBC three studio television cameras are in use. Each of these cameras is capable of getting long shots, medium shots, or close-ups. One of them is mounted on a movable platform, or "dolly," to enable it to make shots while in either forward or backward motion. While one camera is registering a scene the other two cameramen are receiving direction through headphones as to where in the studio they should go and on what objects to focus for their next shots.

The director is seated in a control room where he may view the output of any camera at any time. He, therefore, composes each picture before it goes on the air by relaying his instructions to the cameraman.

How does television direction differ from other forms?

A comparison of technique with that of the stage and the motion pictures would be the most apt way of demonstrating the advantages, differences, limitations, and possibilities of television.

The television director who is experienced in the theater takes advantage of that background, or knowledge, in directing a television show. In many respects direction of the two is the same.

In a dramatic presentation the director selects the actors and they must be the best he can get for the parts to be played. They must look, sound, and give the impression of being the characters they portray. The television director conducts all rehearsals and is responsible for the entire production just as the director is in the theater.

However, neither television cameras nor studios can give the space and size which the stage offers. Consequently, television directors must plan their shows so that only small numbers of people have to be visible at a time. Concentrated action is a keynote in good television direction.

On the other hand, the television director, with three cameras at his disposal, can shift from set to set during

The Director

the action in the same manner as in motion pictures. On the stage, one must either black out or lower a curtain in order to effect a change of setting.

Television also takes advantage of many screen devices and trick-angle shots, close-ups, concentrating on certain bits of business or action in order to heighten a dramatic effect or to clarify a point which may be obscure at a distance.

Both television and the stage have one very great advantage over movies. Both convey action which is happening at the moment. Audiences are impressed with the magic of witnessing an actual event rather than a cold record of what has already taken place. It is the perfectly normal reaction, which will always keep the living theater in existence.

There is a great difference between motion pictures and television direction. Pictures are made in sections, or scenes, or "takes." There are numerous retakes. There is the fabulous cutting room in which, according to many, mediocre acting and stories can be turned into good pictures by adroit editing. But in television, as in the theater, action is continuous, with no allowance for mistakes. It is the director's job to see to it that none are made.

We have found, too, that the television director must set a faster pace for his presentation for the mechanical reason that a person watching a small screen is required to concentrate more intensely. The tempo must be stepped up not only to maintain interest but also to cram the same story into a shorter running time. We have found, for example, that plays should not run much more than an hour, regardless of their quality.

The director also assumes the responsibility of choosing and adapting material from which his shows are made. He must keep in mind that the result of his work will enter homes and the material must be fit for home consumption.

There has been very little original writing for television at NBC since a regular schedule went into effect. That took the form of dramatizing works already existing in other forms—a short story and a novel, to be exact—and creating adaptations of them for television presentation by writing dialogue where needed and, in general, visualizing the story. These adaptations were made by NBC staff directors.

Studio television production at the New York NBC studios has already included almost every conceivable form of entertainment, instruction, and exposition. Prior to May 3, 1939, many experimental broadcasts were presented, including dramas, variety acts, novelties, educational features, and musical ideas. On May 3, NBC offered its first regularly scheduled studio show and it embraced a representation of virtually everything that had been tried experimentally. The content of that show is now, of course, a matter of historic record.

Lowell Thomas, news
Tele-Topics, the first newsreel produced exclusively for
    television
The Three Swifts, jugglers
Richard Rodgers, concert pianist, and Marcy Wescott,
    songs
Millions for Safety, film
Fred Waring and His Pennsylvanians
Mobile-unit pickup from the New York World's Fair,
    interviews
The Unexpected, a play by Aaron Hoffman, with Mar-
    jorie Clarke, Earle Larimore, and David More
Donald's Cousin Gus, a Walt Disney cartoon
Helen Lewis, mistress of ceremonies.

Studio shows have fallen roughly into three broad classifications, and we will discuss each division with regard to the differences in production and direction and presentation.

The Variety Show

This type of show has a distinct advantage over others in many respects. The word "variety" automatically suggests a wide appeal. It is the director's job to engage separate units which when presented in combination will make for the most diverting production.

We Present Television

Let us assume that the producer has booked a dance team, a comedian, an instrumental novelty trio, and a spelling bee. Add a master of ceremonies and the probability is that a good hour's entertainment will result.

There are literally hundreds of possible variety hours. We have produced many successful shows in which very little if any of the talent was professional in the sense of having had experience in vaudeville, musical comedy, or night clubs: spelling bees and quizzes in which people from all walks of life participated; dancing lessons taught by professionals; lantern-slide or otherwise illustrated lectures on art or science or needlework; cooking instruction involving the actual preparation of foods; and many other widely assorted bits and pieces with both eye and ear appeal have been presented by NBC's television cameras with varying success. And, always, it is squarely up to the director to work out his routine and to televise each item in what he believes to be the most effective manner.

Let us suppose that a variety hour opens with the dance team. The use of long over-all camera shots is imperative here in order to show the co-ordination of the dancers—how their entire bodies respond to the rhythm and melody of the music. Frequently, close-ups may be employed to show a particularly intricate bit of footwork or a facial expression. In the main, however, a dance routine must include the full-length shot of the dancers in order to be effective.

We have found we can get a more pleasing effect by elevating the long-shot camera and shooting the dancers from just above the eye level. This increases the illusion of perspective. Bows at the end of the act are usually taken on a medium shot.

The director orders a fade-out or a direct cut to the master of ceremonies, who introduces the next item on the bill.

Let us assume that our comedian is one who works with comic accessories or props. He has a trick hat and collapsible shirt front and that sort of thing. Naturally the director is going to look for items which will make amusing close-ups and thereby heighten the comedy. The comedian also tells funny stories which are augmented by facial pantomime. Here the close-up cameras come in for action. And so it goes.

The hypothetical variety hour continues with an instrumental trio, and here again the director must see to it that the audience gets the greatest possible benefit from their performance. Close-ups showing instrumental detail suggest themselves.

We have learned that audiences like to see what makes the sounds they hear. They demand that television show them just how a musical saw is played, for example. That is television in essence. The audience must feel itself almost a part of the entertainment. Television complies with this by bringing the entertainment into the homes of the audience, at the same time amplifying certain details to make the performance more intimate.

We come now to the last half-hour of our variety show: the spelling bee. The master of ceremonies introduces the master of the bee. (At NBC Paul Wing, who calls himself "The Spelling Master," has successfully filled this role.) Spelling bees are old but we have found that this form of diversion makes excellent television material.

Here, if ever, the audience is taken right into the proceedings. The people competing in the match may be well-known personalities or they may be ribbon clerks. It really doesn't seem to matter a great deal, for spelling is a great equalizer.

The audiences enjoy the spectacle of seeing others struggle over a word and take a not unnatural delight in knowing how to spell a word which is at the moment baffling the person in the studio who has to spell it.

What has the television director to do with a spelling bee? The answer lies in what we have said about audience reactions. The director must follow the progress of the bee, get close-ups, and in general hold the show together just as if it had been carefully rehearsed. One can readily

Television's first mobile television station, used in relaying athletic contests and other outdoor news events.

An interior view of the "flying laboratory" plane showing the lightweight equipment that transmitted the first program from the air.

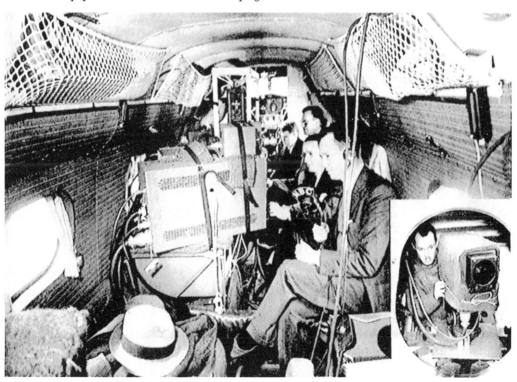

Television in action outside the studio. Bill Allen furnishing the word description of a football game to televiewers;

Dr. Raymond L. Ditmars of the Bronx Zoo exhibiting some of the park's extensive collection of reptiles;

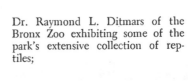

Columbia and Princeton Universities inaugurating the television of baseball, May 17, 1939;

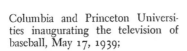

NBC television experts telecasting a fashionable charity ball at the Waldorf-Astoria Hotel.

appreciate the dramatic value of a close-up of some business magnate's face while he is trying to spell "abattoir."

It must be remembered, also, that unrehearsed programs require a great deal of quick thinking on the director's part in order to give a coherent sound-and-sight account. Shows of this type can easily run for thirty minutes without becoming tiresome.

And there you have an hour of variety entertainment which the director has had to keep moving smoothly and smartly. He has also had a great deal to do with arranging the show in advance, which suggests another field of directorial activity which we shall discuss later in this chapter. Here, now, is another great division of television entertainment.

The Comic Opera

Alton Cook, writing in the New York *World-Telegram* of October 19, 1939, said:

"Probably the most successful of the television evenings were the Gilbert and Sullivan presentations. They caught the light tone of mockery that casual Gilbert and Sullivan shows usually miss. The shrewd use of the small stage to which the cameras are still confined was striking. Casts as large as these frequently have seemed crowded uncomfortably into the television scenes."

Since it is common knowledge that newspapers have

not been particularly generous with praise for television, the above excerpt may be considered a high tribute to the form of television show employing music as an integral part of its presentation.

When NBC decided to attempt hour-length shows instead of adhering strictly to variety, Gilbert and Sullivan operas naturally suggested themselves. The first one attempted, *The Pirates of Penzance*, was presented on June 20, 1939, and was an immediate success.

There were several reasons advanced for its enthusiastic reception, chief among them being the inherent entertainment value of Gilbert and Sullivan which has withstood the test of time in the theater and on the radio. However, it represented a radical departure from anything that had been done in television and posed a number of production problems.

The director selected to handle the show was told to confer with Harold Sanford, well-known musical director of light operas in the theater and on the radio, and work out the production with him. Miss Ivy Scott, who has played and directed Gilbert and Sullivan operas for many years, was asked to assist with the intricate stage business which has become traditional with Gilbert and Sullivan.

These three—Mr. Sanford, Miss Scott, and the television director—cut and changed the script and musical score to suit television and assembled a cast. After two weeks' rehearsing, the show went on.

One of the many things we learned from *The Pirates of Penzance* was that not nearly as large a cast was necessary to a televised opera as would be required on the stage. We found that television, with its facility, mobility, and flexibility, could *suggest* the presence of far more people in the ensemble scenes than were actually present.

We used almost the smallest possible orchestra for a very good reason: economy in money and in studio floor space. That, too, was found to be perfectly adequate. The orchestra was never seen and there was always something on the screen to hold the audience's attention *visually*.

The typical comic opera includes a farcical situation enlivened by music—solo, duet, ensemble—and, almost inevitably, a plot complication which could conceivably be tragic but which one instinctively knows will work out happily for the two lovers. *The Pirates of Penzance* was a happy choice, for it contains all those elements and certainly has one of Sullivan's brightest scores and one of Gilbert's most colorful libretti.

And what did the director do to get the most out of this happy combination?

On the assumption that a rapid pace was essential, he cut the script and the score closely. The plot, which, heaven knows, is not extremely complex, was made completely coherent but with very little repetition. One doesn't need to devote much time to establishing characterization in comic opera. All the characters are types and they dress and conduct themselves accordingly, so that they are recognizable at a glance. The hero is obviously the hero; the villain is vehemently the villain; the major general is blatantly a stuffed shirt; and so on. There is little occasion in comic opera for subtleties which require time to portray to the audience.

The first few days of rehearsal were devoted to the singing of the show. Musical Director Sanford worked over the score with principals and chorus with the aid of a pianist.

Then Miss Scott and the television director started with the dialogue, movements, and business of the opera with Mr. Sanford continuing his supervision of the music.

There were many compromises between traditional staging on the part of Miss Scott and the requirements of this new medium on the part of the television director.

Typical conversation:

"She then goes to extreme left stage and he goes in the opposite direction."

"But, Miss Scott, I want them to stay together."

"Oh, it's always been done this way."

"Yes, I know, but this is television."

They stayed together.

The Director

One of the regular costume houses was consulted about costuming *The Pirates of Penzance*. Here, too, he ran afoul of tradition in a few instances, for television does not like solid black or solid white for technical reasons.

As the rehearsals progressed possible camera shots were noted and set with the players.

Let us take time out here to explain that at NBC we have a large rehearsal room in which we indicate the approximate dimensions of our television sets and conduct all preliminary rehearsals.

In the meantime, the television director had collaborated with the scenic designer, with Miss Scott, and with tradition, in sketching the two sets required for *The Pirates of Penzance*.

The special effects department was also requested to build a miniature castle for the purpose of showing an exterior view. In this instance we first showed the castle exterior (miniature) and then cut to action inside the castle (studio set).

It was thought that a commentator would be helpful in setting the scene for the audience and, in general, telling about the cast, the history of the opera, and about the author and the composer. He opened the performance, spoke for about three minutes during intermission, and appeared again at the close of the show to repeat the cast. The idea worked out so well that we used it on our other comic opera telecasts.

In the two camera rehearsal days we worked out all camera positions and shots. We used close-ups on solos, medium shots on duets, and then cut in for close-ups on individual parts, three-quarter shots on larger groups, and long shots to cover mass action and ensemble singing.

We found that on the ensemble numbers it was good to cut into the group with close-ups of principals while the number was in progress to get away from any possible monotony resulting from looking too long at a comparatively large assembly of singers on television's small screen.

We placed the orchestra to one side of the studio with Mr. Sanford in position to conduct the singers as well as his musicians. We built the first act set, "a rocky seashore on the coast of Cornwall," in one end of our rectangular-shaped studio and the second act set, the castle interior, in the other end. In that way we had simply to reverse our camera positions during the intermission.

And so *The Pirates of Penzance* went on the air for the first time with sight and sound.

One cannot refrain from speculating that Gilbert and Sullivan, with all their brilliant imagination, never dreamed of this eventuality.

The Director

The only adverse criticism we heard was that the performance was not long enough, which was actually the most commendable thing that could be said of it.

About a month later we produced *Cox and Box*, a one-act opera which Sullivan wrote in collaboration with F. C. Burnand, following virtually the same directorial technique.

On September 5, 1939, we telecast Gilbert and Sullivan's *H. M. S. Pinafore*, and the experience with the other works showed in a smoother performance.

We found that during the course of a musical number it was wise to change camera shots on a musical beat which calls for close attention to the score. Just such little details go far in tightening a performance and contributing to its progression.

Throughout the entire script of the opera, lyrics included, camera shots were written in, even between words of a phrase, when by switching at that point an effect could be heightened.

Yes, good comic opera is definitely television material which affords entertainment to a wide audience and provides the director with many opportunities for effective presentation. This field of entertainment calls for great care on the part of the director, however, for comic opera demands a brisk pace and effective camera work in addition to clever artists.

We Present Television

The Drama

We come now to the field of television enterprise which demands most of the director. In no other form is so much versatility, originality, and imagination required. Here we find the director in the truest sense; one who interprets a story for the audience.

Particularly in variety shows and to some degree in musical comedies, the director's chief responsibility is to convey the actual entertainment content of the material with which he is working. However, he is allowed far more freedom in plays. That same freedom compels him to apply more discrimination as to what is important; what is to be emphasized; what is to be deleted; what is to be expanded.

Many plays have failed in the theater because of inadequate direction. Conversely, many mediocre plays have become hits because the director has breathed something more into them than the playwright had given. Television, owing to its extreme power of selection and isolation of certain points by means of close-ups and other visual devices, calls for exceedingly precise direction. Everything should be rehearsed carefully and then played exactly as it was rehearsed.

Probably the best exposition of how a play is directed

for television consists in choosing a typical production which was actually presented and going through a more or less detailed description of what the director had to do with it. So, to the best of our recollection, this is how *The Farmer Takes a Wife* was produced at NBC:

Walter D. Edmonds wrote a novel called *Rome Haul*. It told of the people who lived on the famous Erie Canal in upper New York State in the eighteen fifties. The novel was dramatized by Frank B. Elser and Marc Connelly for the theater and it was produced in New York in 1934 with the title changed to *The Farmer Takes a Wife*. Later, it was bought for motion pictures. You may have seen Janet Gaynor and Henry Fonda in the leading roles. We bought the play for television in the fall of 1939. So much for history.

The first thing that had to be done by the director was almost a redramatization of the book. The theater version depended a great deal on big scenes which we knew could not be done in television.

With the mob scenes simplified, there was little of the play left except the basic love story, so we decided to divide the emphasis between the plot and a semidocumentary treatment of the Erie Canal itself. We told our love story but presented it as a legend which grew during the period of the canal's greatest activity. Obviously, this required a considerable amount of rewriting and original writing.

As is explained in other parts of this book, regular motion picture film televises perfectly. We decided, therefore, to shoot special pictures of the canal and use a narrator to tell parts of the legend while the audience was looking at magnificent shots of the canal. The narration included quite a bit of historical material on the period.

We found that there is a canal in Pennsylvania which today looks the way the Erie did in 1850. We took a cameraman and made all the motion picture shots of it we wanted, including an old set of wooden locks opening and closing.

Railroads, which subsequently reduced the canal's importance as a transportation medium in this country, play an important part in the drama, so we wrote a special narrative tying up the railroads' invasion with the story and put that narration with a film montage of old locomotives and trains. This, of course, was actually a film editing job.

The special effects department was called upon to build, in miniature, a section of the canal showing locks opening and closing and actually taking a canal boat through. Overlooking the locks was a replica of a typical hotel of the period complete with a sign reading "Hennessy Hotel." Much of the play's action took place inside the hotel for which an interior set was built in the studio.

In the meantime, the play was being cast and rehearsals had started. Many members of the cast had played their parts in the theater; others were chosen by the director from the vast number of actors who had come in or were sent to see him. Experience in the theater is most important to television acting, but that point is amplified in the part of this book devoted to actors. Let it be stressed here, however, that the ability to sustain a performance—which is required by television—can be learned only in the theater.

The Farmer Takes a Wife still presented technical problems despite its simplified script. It was found that in at least three places in the drama as many as six or seven people had to be in the scene. The director had to establish the presence of that many, but, because of the small picture and lack of studio space, the audience wouldn't have had a very good opportunity to see who was speaking lines. Consequently, we had to arrange certain groupings of the characters so that two were included in the field of one camera and three in another. Switches from camera to camera were made as the dialogue shifted from one part of the group to another. Obviously, all this required careful rehearsals in order to determine the exact positions of the actors.

We Present Television

The scenic designer had his problems, too, with this show. He had to design, with the aid of the director, an interior of an Erie Canal hotel of 1850, a section of the canal wall with a canal boat tied to it and with the front porch of a house giving onto the canal, and the interior of a canal boat cabin. He also had to supervise the construction and painting of these sets in the workshops.

Then a costume house was called in to clothe the cast in the period. Our property men had been sent scurrying after furniture and other required articles.

The director was holding regular rehearsals in which characterizations were worked out with the actors and the entire play was given pace just as would be done in the theater. In rehearsing for a television performance, the director has to keep thinking in terms of possible camera shots in addition to adhering to many of the tenets of theatrical craftsmanship.

Actors trained in the ways of the theater sometimes give the director a strange look when they are told to stand a certain way or do a certain thing. They know, instinctively, that they are being asked to violate some deeply ingrained principle of acting. A gentle reminder that "this is television" is usually enough to quiet their fears.

We have found the use of printed titles effective in introducing regulation credits at the opening of a show and also for indicating intermission periods. Copy for *The Farmer Takes a Wife* titles had been drawn up and placed in the special effects department. It was decided to use a map of the Erie Canal as a background motif and change from the map to our motion pictures of an actual canal as the narrator took up where the printed titles left off. We went from map to film to miniature to studio set and the play.

Background music was thought to be essential to heighten the dramatic, scenic, and historic parts of the play. That was selected from phonograph records which ranged from a group of old Erie Canal boating songs to symphonic excerpts. It should be added that phonograph records are played directly into the sound output of the studio.

The play was replete with sound effects: boat horns, insect noises, train whistles, bells, water sounds, and others. NBC's sound effects department is quite famous in radio, so no major problems arose in this connection.

The director always prepares a shooting script of his entire play before the show goes into actual camera rehearsals. The right hand margin of each page is used for penciled directions, camera shots, orders to cameramen, sound effects, film studio warning and starting cues, music cues, and orders to the stage manager who works in the studio in direct contact with the cast.

We Present Television

A final checkup the day before *The Farmer Takes a Wife* went before the cameras revealed these, up to this point, widely scattered elements: a shooting script, a cast which had been rehearsing for almost two weeks, a set of titles, a narrator, four special film sequences, a miniature, sound effects, marked and timed phonograph records, costumes, properties, and three sets which almost overflowed our floor space, in addition to all basic studio equipment.

It was now up to the director to blend all those factors into a television production.

We devote two days to camera rehearsals on plays at NBC. Every minute is needed. Despite the care the director has used in his preliminary rehearsals, there are always some things which have to be changed because of lack of space or physical inability to produce a shot which seemed right in theory; or it is found that a planned shot is not good pictorially and that a more effective way of establishing some dramatic value must be found. Then, too, the cameramen must become familiar with the show in order to understand such orders as, say, "Number two, get a close-up of Molly when she gets on the steps."

The director works with two screens, or images, before him. Directly facing him is the picture which is going on the air. Slightly to his right is the preview screen, on which he may at all times see what any one camera is

The Director

taking. All shots are first composed on the preview screen, as mentioned above, and then, on cue, the picture is snapped to the "on the air" position. Thus, the television director is—or should be—ahead of the actual performance by at least one camera shot. In other words, while one scene is on the air the next shot, whether it be a totally different scene or another view of the same scene, is being prepared. During all this, the director is trying his best to hold all the other parts of the show together by giving orders to his stage manager through a headphone set, as well as cuing music or film to the engineers in the control room with him.

A stranger, suddenly finding himself in the control booth of NBC's television studio 3-H might be somewhat bewildered at seeing and hearing something like this:

A man whose eyes keep darting from a script on which a spotlight is shining to some electrically reproduced images on the wall before him. The man talks and at times what he says is repeated into a black mouthpiece by another man. Sometimes he talks into a black mouthpiece of his own. Occasionally he lifts his voice to say something to one of the other two men in the room. One can tell at a glance that he is deeply engrossed in what he is doing. At some point a moan may escape from the depths of his being and one is aware that something has made him unhappy. It is even possible that he may suddenly shout or gesticulate wildly, but it doesn't seem to accomplish anything. He soon returns to comparative placidity, though there is no indication that he will remain in that state.

"Take two. One, get on the three men by the door." A quieter voice repeats the last command. "Three on the girl." Repeated. "Two minutes, 5-A." Repeated by a different voice.

"Get that boat horn further away from the mike. . . . Preview one. . . . Warning on your music. . . . Take one. . . . Two, get on title.

"Noel, sneak in when we're off that side of the set and get that lamp chimney. It's making a bloom. . . . Take three. . . . One, stay there. . . . Take one. . . . Three, get on the miniature." Repeated. "One, dolly in." Repeated. "One minute, 5-A." Repeated.

"Oh, no, two, no, on the title!" Repeated but without so much intensity. "Hold it, one." Repeated. "Boat horn. . . . Sneak in your music. . . . Thirty seconds, 5-A." Last part repeated. "Preview three. O.K. Roll the film, 5-A!" "Roll the film, 5-A" is repeated. "Stand by, narrator. . . . Fade out one, fade in five. . . . Music up. . . . Preview three. . . . Music down. . . . Start your narrator. . . . Quiet in the studio! . . . Take three. . . . Preview two. . . . Music up. . . . Take

Television, as indicated by these Don Lee programs on the West Coast, has wide appeal for people of all ages with a great variety of interests. A model building program for hobbyists with Reginald Denney in Hollywood;

a home economics demonstration— housewives see how a famous chef mixes a salad;

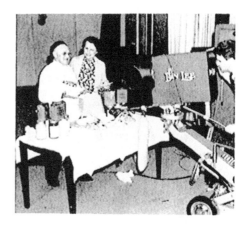

an art appreciation program by Arthur Baker on sculpture.

The Tournament of Roses Parade at Pasadena is telecast by the Don Lee portable transmitting equipment.

RCA home facsimile receiver. (COURTESY RCA)

Home facsimile receiver showing the stylus used to scan the prepared paper. (COURTESY FINCH TELECOMMUNICATIONS)

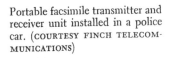

Portable facsimile transmitter and receiver unit installed in a police car. (COURTESY FINCH TELECOMMUNICATIONS)

two. . . . Kill the studio sound. . . . O.K., intermission, three minutes."

That man was a television director and the stranger had seen and heard a part of a television performance. The fact that a loud-speaker in the same room was giving forth the dialogue of a play with attendant sound effects and music only helped to confuse the stranger. But if the director felt any confusion, he concealed it. In fact, he appeared far less perplexed than he did that night in the Players Club bar when somebody asked him, "How do you direct television?" and he had answered, "Oh, you just do, I suppose."

CHAPTER SIX

The Actor

BY EARLE LARIMORE

ASK any actor how he would like to be in television. "Suppose I forget my lines," will be his first question. "How am I going to get cued?"

The sensitive television microphone makes prompting an open and embarrassing secret. We cannot read from scripts as in radio nor can we cut a scene as in the movies, nor can we receive a muffled prompt from the stage manager offstage as in the theater. The play is all telecast as we stand before the camera and deliver our lines to you. But like most fears, our dread of "going up" in our lines is seldom realized, so cuing is not one of television's most pressing problems. It's an interesting question, however, because it is typical of the many new details in television that puzzle the actor.

Picture the poor actor at his first television rehearsal! He has come directly from the stage to this new picture and sound theater of the air. Everything is bewildering. The lights, strung in multiple banks over his head, dazzle his eyes and make his skin smart. He perspires a great deal. Some actors have even taken saline drinks during their first television rehearsals to counteract any disagreeable effects from the lights. But television lighting does not work so severe a hardship on the actor as such extreme precautions would indicate. The dazzle of footlights is the very stuff of our world. Many of us have acted under the glare of klieg lights in the movies. It is only the first blinking and smarting under the television lights that is dismaying to the actor, especially since he has so many other adjustments to make. Water-cooled lights have been perfected and perhaps will be used in the future to save the actors and the studio staff from any discomfort.

The television cameras are rather baffling at first. There are three of them at present, huge cameras that swing slowly around. When I am on the television set, I think of those cameras as three octopuses with little green eyes blinking on and off, their silvery forms moving ponderously. The green light is a signal that the camera is in operation. One camera angles and takes a shot and another picks it up from a different angle. I was very conscious of this new technical apparatus at first. But I soon learned that the cameras are the director's worry, and the actor has to concentrate on his lines and his "business" much as if he were on the stage. His only concern with the camera is to keep within range. This, too, is something new, even for an actor who has had Hollywood experience.

The television range is small, and this factor greatly controls the way we play our parts and the kind of gestures we use. It also means that two of us doing a scene together must speak our most dramatic lines so close to each other that we cannot possibly deliver our lines in the same way or make the same gestures as we would on the stage or in the movies.

Suppose a lovers' quarrel is taking place on the television screen. The two actors will have their heads nearly together. They cannot show anger by striding away from each other impatiently as they would on the stage. This limitation makes new demands on the actor's flexibility and resourcefulness. He must be able to express tense emotional climaxes in a small space and get his point across no matter how cramped he feels. "Playing close" all the time puts the actor off his emotional stride, but he will undoubtedly adapt himself to what amounts to a new style of acting, a new technique for conveying emotion. Just to give you an idea of what we mean, let us take the simple act of lighting a cigarette as an actor might do it on the stage, the screen, and on television.

The manner in which an actor lights a cigarette, as you know, is often used to express emotion—tension, contentment, impatience. On the stage, we see the actor picking up his cigarette from a box on a table and lighting it with a wide gesture. He can incorporate into that gesture all the subtleties of which he is capable as long as it has a basic sweep. The stage, for obvious reasons, demands the larger gestures. In the movies, much detailed motion may be used. The actor can fuss with the cigarette as he talks, holding it at breast level where the camera can catch him even in a close-up. For instance, he can betray his nervousness by small fidgety movements. The motion picture camera is best suited for recording tiny motions as an adjunct to the portrayal of emotion. When we come to television we find a paradox of acting technique. The actor certainly cannot make so wide a gesture as on the stage in reaching for his cigarette since he easily gets out of camera range. Yet the motion must have definite sweep and largeness to it or the camera will not record the impression he wishes to give. So you see what a strange dilemma confronts the television actor at his first rehearsal. His gestures must not be too small, neither must they be so large as on the stage, yet they must combine some of the features of both techniques. Nor can an actor simply patch up a few tricks from both stage and screen. He has to mold a new technique to fit a new range, and he has to learn to express the same old emotions in ways suited to the new demands of television drama.

Not only must the actor learn to modify his gestures to suit new spatial and recording ranges, but he must accustom himself physically to working in a very small space for the entire length of a drama. An actor calls this "being held in a tight place." Literally, he is. Motion picture actors sometimes work within a limited compass, of course, but they are only "held" there for a few minutes during a "take." Long movie shots such as occur in outdoor scenes give the screen actor much more freedom of movement than the television actor can hope for at the present stage of television development. Television transmission of a play is simultaneous with its acting. It's an actor-to-camera-to-you proposition, and the actors are "held" within camera range until the play is over. This is really difficult for an actor to get accustomed to when he has had a whole stage to stride across in anger or exultation.

The actor's "business" is modified considerably by television. By "business," of course, we mean all those gestures of the hands, sudden turns, widening of the eyes, and so forth that are an actor's emotional shorthand. A great deal of business that is splendid in the movies and on the stage is just a total loss on television. Recently I was rehearsing a television play with a screen actress.

The Actor

She came to a part where she was supposed to be frightened. She stood rooted to the spot and widened her eyes in a very convincing expression of terror.

"No, no," the director cried, "not like that. This is television!" and he showed her how to register fright for the television camera by expressing it in her whole body as she took a startled backward step. The gesture registered beautifully on the screen. The actress had to be careful to limit it more than she would have on the stage, where she would probably have taken several steps backward. In this case, stage technique was more suitable than motion picture technique to convey the mood the director wanted to register on the television screen, but this is not always so.

There is very little facial business used in television at present. Large gestures that remain within the camera scope are most in favor. There is an emphasis on playing with the whole body more than with the detailed facial expressions that the movies have developed in a great many of our best actors and actresses.

Most important instrument of emotion on television is the actor's voice; the director depends to a great extent for his emotional effects on the timbre of the actor's voice and its capacity for conveying a wide range of emotional stresses. The voice, of course, is solidly bulwarked by the visual excitement of having attention fixed on the players.

We Present Television

The actor, when you come to think of it, has had to be pretty nimble to adapt himself to the changes that have taken place so rapidly in his field. Back in Shakespeare's time we find him performing in a circular theater minus scenery and minus mood-creating lights. Gradually the stage became focused from one angle, with the audience in place of a fourth wall. This state of affairs has existed long enough for the actor to become thoroughly accustomed to it. The movies and the radio taught the most flexible of the actors to accustom themselves to playing before no audience at all, as far as the eye could see. Now television makes one more demand on the actor's imagination and adaptability. As far as he can see, in television his audience consists of three cameras! And even more mystifying, the actor does not always know which camera is angling him onto the television screen where his audience views him. In this new medium, the actor does not "play" to any one of the cameras, since the cameras are the directors' province.

It was a pleasant surprise to me and to other actors to learn that in television I could turn my back to the audience without fear of muffling my voice. The moving microphone takes care of that. So you see the television stage gives the actor new freedom of movement in one direction while it restricts him in another. It gives him

The Actor

the opportunity to experiment with groupings of people that are not practical on the stage.

Perhaps I have laid too great stress on the features of television that temporarily hamper an actor. If I have, it is because actors, like everyone else, notice the things most that joggle them out of the old accustomed routine. At first we don't see these features as a chance to project ourselves into a new technique with plenty of room for expression if we have the imagination to take advantage of it. We're disturbed, as anyone would be, at having so many new adjustments to make. It is only as the advantages of television really dawn on us that we accept the new medium with enthusiasm and realize its full challenge.

The greatest feature television offers the actor is *intimacy* with his audience. Intimacy is, of course, the one quality that every actor strives for above all others. He wants you to feel that you are doing a bit of eavesdropping on the life of the character he is creating at a moment which puts that character to a crucial test. The more an actor can convey the impression that he is letting his audience in on a secret the more successful he is as an actor. Imagine the actor's excitement when he first "walks" his part on the television stage! He feels that all the technical discomfort of television is trivial compared to this one magnificent feature.

We Present Television

We know that when our play is ready to telecast, we shall walk right into your home. We shall appeal to your emotions as you sit relaxed in your favorite easy chair before your television screen. We play the play through to its climax with all the emotional unity that that implies. We do not have to do a middle scene last and a last scene first as happens often in the movies. We do not have to worry, as we do in the theater, about the man in the last row in the balcony who may be our most sympathetic spectator and yet it's almost impossible for us to give him as good a performance as the man in evening clothes in third row center. We can throw aside all these cares and act for you as we have never acted before. We unfold the story; the pleasant living room of the viewer is made magic with the characters on the screen, with the things they say and do.

What medium has ever offered such a happy hunting ground for the actor? It is our dream of contact with many people of all classes and races and occupations under the most intimate conditions ever offered for actor or audience.

You can see how thrilled we are with the possibilities of television. Radio gave us the thrill of an almost inestimably large audience composed of many different types of people. But the voice is only part of our register of emotion. Television had to come to round out

The Actor

the picture and yet retain and even broaden that great audience of people of all ages and tastes and occupations. Now the actor may talk and be seen. Only we actors can fully realize how great is the fixative power of being seen, of conveying in one significant gesture which is the height of our art something that a thousand words can never say so well. Now truly, in television, we have our ultimate wish and we are eager to give it our best.

Actors expect a great deal from television. We have a definite attitude toward it that we hope will be justified. When anyone tentatively accuses television of not giving its best, the stock excuse is that it is still in the experimental stage. But is it? Certainly, it is no longer in an experimental stage technically. Then what do they mean by an experimental stage? Do they mean that they have not discovered yet what type of program is best suited to television broadcasting? We certainly welcome the sort of experimentation that will develop better or new types of programs on television. It is only in these newer mediums that the actor can be presented to the television audience at his best. Therefore, he has a very real interest in the subject. Surely "experimentation" cannot be used as an excuse for presenting only safe and sure stuff that has filled in as program material from the early days of the movies right through radio. Torch singers, swing bands, variety acts, humorous sketches, mystery plays, all these have their place in television. But we hope television is not going to stop at that. As actors in close contact with the public, we feel that television underestimates that public if it thinks the presentation of new types of programs is too daring a venture.

The actor knows the American public is singularly receptive to new ideas. Already it is asking eagerly about television, figuring costs for home receivers, buying sets. One grave danger suggests itself, the danger that television will be just a fad with the American people. Television certainly must come. It is needed; it is the ultimate in communication between human beings. But there is serious danger that it might be greatly delayed in public acceptance by mistaken handling of program material. The actor has a personal interest in avoiding that catastrophe, for he has an economic as well as an artistic stake in television. Television means tremendous opportunities for employment among actors. It will ease, if not solve, their ever-present economic problem. Television will use many more actors than stage, radio, or screen. It cannot use the same actors over and over on different sketches during one day, as in radio. The actors will be visible; the audience will demand a variety of faces and types. Several television plays will be rehearsed a week with much longer rehearsals than in radio because of the impossibility of using scripts in television broadcasting.

The Actor

The actor has an interest in finding out what sorts of programs are best suited to television broadcasting and what the people will applaud. The problem is to discover what television can do that *no other medium can do*. The actor cannot solve that problem. It is a relationship primarily between the directors, the scriptwriters and adapters, and the public. But the actor knows that once television can offer a type of entertainment that only it can produce in the most artistic and satisfying manner, television is here to stay. And the actor wants to see that happen.

What is this about television being just a fad? Picture a man who has just purchased a television set for his home. He has invested quite a large sum. He has put himself ahead of his neighbors by his foresight in being the first to install a set in his home. Now he stretches out in his easy chair. With a glow of anticipation he switches on the dials. A perfect image soon takes form on the screen. If television offers him no different programs from those he has got on his radio, why, he asks himself, did he spend all that extra money?

This kind of disappointment is likely to develop into resentment against television itself. It isn't the man's fault. It isn't television's fault because the apparatus is mechanically perfect enough even at this stage to satisfy anyone. It's a thing for the television men to keep in mind even if they have to work their imaginations over-time to create programs that are exclusively and excit-ingly *television* entertainment.

This certainly is not easy. It is always easier to adapt to one medium what has succeeded in another. The early talkies suggest themselves in this connection. They pre-sented literal reproductions of successful Broadway plays. But the movies only became a great medium and a great entertainment field when the directors discovered that the camera had its own dynamic language. Then movies grew up. So did motion picture actors. They commanded new respect in the public eye when it was discovered they could present a story in a new way equally as exciting and artistic as Broadway's technique. The public liked the change. It gave them tremendous breadth, telescoped action. It gave them picture stories that no longer leaned, like a cripple who has outgrown his crutch, on stage talk and mannerisms. It took the movies a long time to tum-ble to this. But television will not have so long a period of grace to realize its potentialities. Either it clicks now or it goes on the shelf for another ten or twelve years, when some group willing to listen to the people's ideas of what they like take it up again and do the thing right.

Of course, you'll ask us: If you're so smart, what kind of programs would you suggest for television? We are not directors. But as actors who know the public and who have had experience in other entertainment mediums, we do have some ideas.

We think first that television should feature the drama that depends upon good writing for its success or failure as a play. It should feature the plays that depend upon thought for their point rather than a great deal of action. To my mind, Shakespeare is one of the best bets for television because his plays embody the very spirit and space requirements that are most suited to television drama. Shakespeare's plays are intensely dramatic but their drama is in the thought and emotion back of the action. The girl being untied from the railroad track by the hero just as the train approaches is never the point of a Shakespearian play. The point is the beautifully expressed thoughts and attitudes of the characters in relation to the action and to each other. Action is incidental to thought. Television, with its small range and intensely intimate focus, cries out for this type of material. This kind of play is always most successful when a small group is concentrating upon it. That, of course, is exactly the type of audience television reaches in the home.

Approaching modern times, we find an increasing trend toward this emphasis on thought over action, as in the plays of A. A. Milne and Eugene O'Neill. *Berkeley Square*, by John Balderstone, would lend itself beautifully to television presentation because it embodies the qualities we have mentioned. Another ideal play is *Outward Bound*, by Sutton Vane, in which each scene has a thought to express. Each scene has something to say instead of something to *do*.

Why bother with television, then, you may ask, if speech and its quality is the important thing? Why not stick to radio? Actor and critic agree on the answer. *Without the visual element, you simply cannot get the ultimate in beautiful thought.* You must have the visual appeal to fix the attention of your audience and to prepare the mind for what is to come. That visual attention once arrested, the receptive mind will do the rest. To emphasize this, think how many times your attention has been diverted from the radio, and how easily, even during a thrilling piece.

These ideas of what constitutes the best material for television drama are by no means an actor's daydreaming. Among the most enjoyed features television has put on were the Gilbert and Sullivan comic operas. The clever librettos with their plays on words and their social satire appeal to a very civilized sense of humor. They appeared to perfect advantage over television with their action and brilliant eye-fixing costumes subordinate to the gay music and the intimate style of acting and singing.

So you see this feeling the actor has about the suitability of plays like those of Shakespeare, of Milne and

The Actor

O'Neill, is no actor's theory but an established fact. We hope the directors will apply this knowledge to the presentation of distinctive television programs. Audiences have shown that they like this program material. We want more of it because it gives us a better chance of having our audiences like us.

We actors are very excited about the dramatic possibilities of the mobile unit. Here is a brand new challenge to our best and most artistic efforts. What can the mobile unit do besides telecast football games, parades, catastrophes? It can give us the most exciting new opportunities for acting we have been offered in a long, long time. Plans are already being formulated for doing outdoor dramas with natural scenic backgrounds or even industrial backgrounds. A production of Shakespeare's *Midsummer Night's Dream* immediately suggests itself as fascinating material for this experimental technique.

We see a new technique being developed in the entertainment field by employing the mobile unit to its fullest possibilities. Just as there was a renaissance of French art when the painters went outside of their studios for inspiration, so a new technique may be developed in television by looking for material outside the studios. A new type of documentary play somewhat similar to the documentary film can be developed. Scenarios can be planned with a basis in real life, in the life of the streets, in the insides of factories and homes. This definitely includes the actor. Even in documentary films, the latest trend is to include experienced actors as the focal point of the cast. They will be used in documentary television plays, too. Besides fine plays set against natural backgrounds, the mobile unit can present stories against a living background of the American people, at work and at play, with the actors carrying the story and the people supplying a fluid and dynamic new kind of atmosphere.

Another type of program the mobile unit could develop that would afford opportunities for actors might be a "Know Your America" series in which bits of history and folklore would be re-enacted on the historic spots where they originally occurred. There are endless possibilities. We have barely suggested a few of them. But we hope the mobile unit will take its full place in the scheme of television program development.

Television will have no difficulty selling sets and getting audiences if programs such as these are developed, experimental and awkward as they may be at first. Television will then have achieved the goal it must set for itself: that of presenting programs so new, so different that no medium can compete on the same ground. If Joe Jones is getting these unique features on his television set, you may be sure his neighbor will want one, too. He'll realize very soon that he's missing something!

The Actor

To speak in broad general terms, actors see television as vast new audiences. Like the proverbial pebble thrown into the brook, the present small beginnings of television will widen out to ever-greater audience range, reaching people the actor had only dreamed of including among his audience.

The actor is thrilled at these possibilities. Broadway theater, because of space and price limitations, has perforce confined the audience to a very small group of people. The movies gave the actor a taste of the vast hinterland that legitimate acting scarcely ever touched except on rare road tours, and even these appealed to people with a certain minimum income. Radio gave the actor only a voice; it robbed him of his carefully developed stage technique. Neither medium gave him the satisfaction of his beloved stage, where he can create a three-dimensional man that his audience can grow to love and never forget. Now here is television with its promise of vast audiences and its opportunity for sustained emotional performances. The actor can not only be seen but he has the advantage of playing his part from the beginning to the end of the play as on the stage. Best of all, he can be seen by a great audience of people in their homes. This is the fulfillment of the actor's vision of the vastness of audience commanded by radio and motion pictures, combined with the sustained and unified performances

of the stage. Can you wonder that actors everywhere regard television as another actor's dream come true?

The actor wants to do big things in television. He
knows the public. He has great respect for its intelligence,
for its eager response to superior entertainment whenever
it is presented. The actor believes television can be one
of the most successful entertainment mediums ever developed *if* the people are allowed to express their *real*
preferences. Television belongs to the public. They must
tell us what they want.

CHAPTER SEVEN

The New Newsreel

BY CHARLES E. BUTTERFIELD
Noted Radio Writer

THERE'S more magic in the air waves nowadays. Think of seeing via the ether, seeing so well that "spellbinding" is the only word for it. But for the future, greater things are in store. Someday, it may be a commonplace to switch on the home receiver and get a lifesize picture in vivid colors on the living room wall, fortified with true sound reproduction and even accompanied by the pleasing aroma of a springtime flower garden.

This may be carrying a prophecy too far; nevertheless enough has become an actuality to indicate that the complete fulfillment will be less of an impossibility than it sounds. Television is bringing forth a new newsreel—in fact, *the* new newsreel. It is a newsreel unlike that of the movies, unlike anything attempted heretofore. It is so different that ultimately it may not even be regarded as a newsreel, but more as the firsthand copy of an actual event as it takes place. It had its vague beginnings in

369

broadcasting, with its vocal descriptions of sporting and other activities. That beginning attempted to convey by ear appeal the full effect that only sight and sound together can produce. This has always seemed to me like the plight of a nearsighted man at the movies without his glasses. He can hear excellently what is going on, but when the dialogue switches to the visuality of the screen he is bound to miss something in important detail, just as the radio listener does when depending upon the announcer's voice to give him a complete picture of a football game.

What the new newsreel ultimately is to be titled—newsreel doesn't quite seem to fit, for there is nothing like the reel of the films—may depend on the quick wit of some word inventor. Here are a few terms that could be applied, any one of which might do in one or several instances: televiews, telescenics, teletopics, televents, teletime, more probably telenews.

Televising out-of-the-studio events has been made possible by the television engineer, who, taking a page from the history of his sound broadcast predecessors, has made part of his equipment portable. By mounting a special transmitter and attendant equipment on wheels or putting it all in suitable carrying cases, he has made it feasible to go afield and catch a scene as it unfolds as easily as the cameras are turned on a prepared drama in the studio.

Thus, just as in sound broadcasting, the mobile apparatus is taken to a spot close enough to the main transmitter so that it can relay the sound and sight of a particular event back for general broadcasting. Along goes an announcer, for he still has a story to tell in telecasting, though without the wordiness of a sound program; also the cameramen.

The cameramen are the important additions. Upon them falls the job of getting the pictures for a lookable program. As many as three cameras can be hooked in so that no part of a scene is missed. In turn, they take close-ups, distant views, and medium shots, working a little after the technique of the movie camera, except that there can be no retakes. The scenes then go right on the air.

A crew of engineers operates the relay transmitter, handles the field unit controls, and does the other jobs needed in the pickup of open-air television. There's hardly an event that takes place outdoors, or indoors for that matter, which cannot be turned into images by the mobile equipment.

Sports events lend themselves most readily to the tele-camera because of their continuous action. Since they are scheduled in advance, preparations can be made ahead of time to cover them with pictures through the air. It is these events known beforehand that are more easily handled with a microphone than are unexpected happenings. The same is true to a greater extent in television, where the technical setup is considerably more involved than that for sound alone.

Thus, it can be seen that the new newsreel is going to have all of the limitations of the microphone, if not more, for some years to come. It will be right on the spot with events classified as "set" in newspaper parlance, but will have to depend on the motion picture newsreel camera or even the spoken word, augmented with such visual elements as can be devised from maps, still pictures, or other devices in instances where direct pickup on the scene will be either impossible or impracticable.

Handicaps or not—and even the movie newsreels have them—when television technique has passed through a long period of ironing out, and just what is and is not good camera material is determined, it will be possible to do things that today seem impracticable.

Spontaneously telecast outdoor scenes have come to attention on occasions. While those cited occurred during test periods of NBC's mobile unit, they could happen in a regular transmission. There is the now famous suicide leap recorded by an air camera as it was picking up scenes in the Plaza of Rockefeller Center and sending them to a special receiver in the television studios of NBC. Again, when the mobile unit was afield for further testing, it televised a fire on Ward's Island, East River, New York, and sent such good images to the receiver several miles away that photographs of buildings ablaze were made from the set and printed in the newspapers.

These two instances since have gone on the record as television "firsts" in spot news. While they didn't go beyond a test receiver, they did prove that if a "break" comes when the telecamera is there, it won't have any difficulty in recording the event.

But the mobile unit is not confined to outdoor scenes. It can park some of its equipment in the street and take cameras and the other apparatus into a building to get indoor scenes. Outdoors it depends on sunlight, or just daylight; indoors it may provide its own lighting effects to augment those already available.

Outdoor scenes put on the air since regular telecasting was started in New York City have not been confined to sporting events by any means. But before going into the field of other activities, let us look over the sports arena to see just what has been tried.

The first such event, as a regularly scheduled program, was a baseball game, the college contest between Columbia and Princeton, on May 17, 1939. A single camera was so set up along the third-base line that it gave a general view of the action. This telecast revealed that one camera was not enough. By the time another important baseball game went into the television records, a professional double-header between the National Leaguers, Cincinnati and Brooklyn, in Brooklyn, a second camera had been added with resulting better relay of the game. Tennis, football in the fall, swimming contests in outdoor pools, and track meets on New York's Randall's Island have come under the eye of the camera, even when the day was cloudy.

Indoor events have offered, among other items, six-day bicycle racing, boxing, and wrestling. Wrestling or boxing is probably the high spot of indoor sports, since it concentrates plenty of action within a comparatively small space. Cameramen quickly learned to set their lenses so that little except the wrestlers or the boxers occupied the screen. In New York's first winter of television, the wrestling and boxing matches were made regular weekly features because of the favorable response from the earliest televiewers. Many of them came from the Ridgewood Grove arena in Brooklyn. While the contestants for the most part were not big-timers, they gave the viewers all of the thrills that championship frays offer.

The huge Madison Square Garden arena also contributed early to the television scene. In fact, it registered a "first" in boxing when the heavyweights, Max Baer and

Lou Nova, performed on June 1, 1939, as the television cameras recorded the flying fists and the downfall of Baer before an estimated audience of 15,000. Every available receiver in town had its crowds of viewers.

But the first winter of television didn't confine itself to boxing and wrestling as indoor contributions. Before spring had come such events at the Garden as track meets, basketball games, and hockey matches had all been televised.

Turning to football, attention was given to both college and professional games available in the New York area. As long as there was sufficient daylight for the cameras, everything that happened on the field, except a flying punt or a tumbling drop kick, could be clearly seen in the receivers. Late in the season, with the days growing shorter, early sign-off from the gridiron was often necessary because of darkness.

Lighting provided by daylight, even when not augmented by the brilliancy of the sun, is sufficient to give good images. In fact, television pictures have been obtained on occasions when movie cameramen could not roll their machines owing to the darkness, or a rainstorm. However, in the case of interior shots, particularly at the ringside, the available lighting has not been so strong as might be desired. The over-all result has been images in which detail has tended to haze somewhat. Still, the action, particularly in wrestling, with the contestants on the mat against its white background, generally has overshadowed the defects.

Right here it might be said that as the art goes forward, engineering knowledge will overcome these handicaps with more sensitive camera tubes. Each advancement in camera improvement makes possible a corresponding reduction in the amount of light required.

Besides the sports mentioned, others have been put on view. Almost any kind of athletic event, from skiing in winter to polo in summer, can be shown in this way. Up to the time the second World War had put a halt to television activities in England, horse racing had been given considerable attention there, though little if any in the New York telecasts. In fact, a horse race was the first outdoor sport, as far as the records show, ever to be recorded by television. It was the famous Derby at Epsom Downs, England.

It is to be noted that the announcing technique is considerably different with pictures than for sound alone. Words are not so important where the action itself is visible. Thus the announcer becomes a sort of spare wheel, his job being to give a touch here and there to stress some point that might not be very clear. In the earlier sports telecasts, continuous description on the order of sound broadcasting was tried. This proved rather unpleasing as the verbal description was too far behind the action. Experience has shown that the announcer must be an adjunct rather than a principal in sports telecasting.

In outdoor telecasting, a variety of subjects was made available, several of them notable, others just ordinary occurrences. Here again, events scheduled in advance predominated. Prime among them was the opening of the 1939 New York World's Fair. This telecast showed the face of President Roosevelt on the air for the first time, and provided for those lucky individuals having receivers a look-in on the start of NBC's regular schedule of programs with the RCA system of transmission.

While Mr. Roosevelt was the first president to be featured in such a telecast, he had a predecessor. Twelve years previous to this, former President Herbert Hoover, then Secretary of Commerce, participated in the test of an earlier system of television. His features were transmitted by wire from Washington in 1927 and were viewed by a special group in New York in a test conducted by engineers of the Bell Telephone Laboratories.

England's king and queen, on their June 10, 1939, visit to the New York World's Fair, were the subject of another high spot in the first year of outdoor television. At one point, the royal couple approached so close to the camera that they almost filled the screen. The satisfactory reception of this program over an airline distance of 130 miles from the transmitter to Schenectady was a record up to that time. A special antenna was used, but the receiver was a stock model.

These are only a few of the special events that have contributed in the opening year to the formation of the new newsreel. Naturally, when television goes afield it cannot do as does the movie newsreel—cover a whole group of events at one showing. When it visits a certain scene, that is the scene upon which it must concentrate. However, duplication of mobile units will take care of that problem.

Airplanes and airports, particularly New York's new LaGuardia Field, have received considerable attention from the mobile unit. Machines landing after transcontinental flights and departing for Chicago and points west, and other activities around a busy sky terminal, all have come within the scope of the television eye.

In airplane telecasting there was the thrill afforded a group of newspapermen who, upon returning from a flight over Washington, where they saw pictures sent from New York while 20,000 feet up, watched their plane land at LaGuardia Field on the screen of the special receiver installed on board. But that isn't all. The airplane itself has been turned into a mobile unit, with a new type of portable transmitter aboard and with a camera pointed out of a special porthole to televise scenes far below. What an advantage that might turn out to be in a military maneuver!

Parades of various descriptions have been made the subjects of other telecasts. One of the most notable was the 1939 Thanksgiving Day Parade of Macy's, large New York department store. Towering balloons depicting the delightful characters of childhood stories were not too big to fit into the screen if they did not get too close to the camera. The result was one of the best outdoor transmissions up to that time.

The man-in-the-street interviews, which have been so popular in sound broadcasting, are considerably more effective in television, making the character immediate. Long before regular programs were started, this type of telecast was given tryouts during test periods of the mobile unit. They were continued from time to time in regular telecasting. Various ideas have been included, but the basis is the man-in-the-street formula. Even other types of outdoor pickups often incorporate the same technique of interviewing.

Broadway's opening of the four-hour movie, *Gone With the Wind*, gave a novel touch to these experiments. This was outdoor telecasting at night. There were scenes under the marquee entrance, scenes of the crowds under the spotlights, of the unloading taxicabs, and of the gaping admirers there to watch the celebrities. Transmission was from the front of the Major Bowes-managed Capitol Theatre, with an additional camera placed inside the lobby for interviews of the celebrities who came to attend an opening that rivaled Hollywood's best. Thus, this was a combination of both outdoor and indoor television handled by the mobile transmitter. It can be contrasted with another instance where all of the action took place indoors. This was at the Waldorf-Astoria Hotel, an hour of entertainment at the first annual Television Ball. The program included a panoramic view of the guests, taken with the aid of a moving spotlight.

Before the regular schedule of programs was started by the National Broadcasting Company, probably the most notable of the indoor pickups with mobile equipment was the televising of the busy New York headquarters of The Associated Press. This was done in mid-April, 1939, as part of a special test program sent through the air to twenty receivers installed in the Waldorf-Astoria Hotel for the members of the AP at their annual meeting.

The feat was accomplished by parking the mobile equipment on the street in front of The Associated Press Building in Rockefeller Center and stringing up cables to connect with the camera installed in the main news room four floors above. For ten minutes the camera recorded the principal functions of news on the move via automatic printers, wirephoto, and the other modern means used to make current events available to the news-papers and the reading public in the speediest manner possible.

It was the first time on record that a telecast had been made under conditions which did not interrupt an important service, and it required more than the usual preliminary preparations to make it possible.

In its weekly schedules, NBC has sought to include at least one and often several programs in which the mobile equipment is used. During some weeks, the studio vied with the mobile unit for honors in originating the majority of the broadcasts. Even ice skating in the Rockefeller Center artificial rink made an occasional appearance on the midwinter screen.

Opinions might differ as to the type of feature which wins the most attention from the viewers. Outdoors it might be baseball, the nation's prime sport, with football a close second. Indoors, boxing should have a high place. I myself would vote for wrestling. It certainly has much more comedy than boxing.

The first year of outdoor telecasting has had its many problems, all of which are being gradually overcome. There were times when electrical interference experienced on the relay channel to the transmitter caused defects in the images. At others, trouble with the apparatus itself plagued the engineers. But all in all, the results have been such as to compensate for the difficulties that have sprung up. Sharpness of the images from the mobile unit has not always compared favorably with those originating in the studios, where lighting and other conditions can be controlled to a greater extent. Also experience has taught that placement of the camera has a great deal to do with the effectiveness of scene reproduction. For instance, continued transmission of boxing and wrestling matches has demonstrated that in cases where the camera takes in only half of the ring—just enough to allow space for the two contestants and the referee—the viewer's view is as good as that from a ringside seat. Sometimes it could be considered as better, since there is nothing to distract the attention, such as jostling crowds.

All of this certainly indicates that a year of television, at least as far as New York City is concerned, has brought about progress in the technique of making things visual through the air. Most of the relay points have been within a ten-mile radius of the main transmitter, although during the Eastern Grass Courts tennis matches at Rye, a suburb of New York City, in August, 1939, the equipment was as far away as 23 miles. The quality of the signals from there was just as good as from points closer.

While close-ups are always more satisfactory because

of the smallness of the receiving screen, distant views register with sufficient detail. Panoramic scanning also comes over satisfactorily. In addition to radio relay, both test and regular mobile transmissions have been made over an ordinary telephone line using special terminal equipment. Scenes of a six-day bicycle race at Madison Square Garden were handled that way in the first trial. Subsequent pickups from the Garden also were sent to the main transmitter in the same manner.

It is the telemobile station that has made all this out-of-the-studio telecasting possible. Actually, it comprises two telemobile units. This equipment made up America's first remote-control television system. Since being put in operation as part of NBC's picture setup, new and improved mobile apparatus, requiring one-sixth the space, has been made ready for incorporation in the television lineup.

Before considering the advanced equipment, let us examine this television station on wheels. Its development was another of the accomplishments of the engineers of the research department of the Radio Corporation of America. It has done yeoman service for NBC, and the job it has done has more than paid for all the trials and tribulations.

The telemobile units make an impressive appearance on the highway. The size of the average motorbus, their exterior coloring is in two tones, silver and blue. In place of the overstuffed seats of the motorbus, the cars carry rack after rack of apparatus—everything that is required to handle a sound-and-sight program in the field. When in operation, a crew of ten engineers is carried. One car has the video or picture apparatus, while the other houses the ultra-short-wave transmitter which relays the program to the main transmitter. The video van has a platform on top where both camera and long-distance pickup microphone can be placed to televise parades and the like. The interior is the counterpart of the studio control room. It contains all of the amplifiers and the many tubes needed to handle sound and sight as electrical energy. One or more microphones can be used, while the picture setup is such that it can pass on the image of two cameras. When the mobile station was first put in operation, it had only one camera, but after a month or so the second was added to give a variety of pickup range.

All of the equipment is mounted on racks extending down the center of the van, affording easy access for repairs or alterations. In the control room section are receiving tubes so that the control engineer can see what is going out on the air and at the same time know what the second camera is recording. In this manner it can be determined at a glance which camera has the appropriate scene. As a rule, one is used for close-ups and the other for more distant views. Besides the radio and television equipment, there is an elaborate telephone cue circuit to keep each member of the operative crew in touch with the directors of the unit.

The cameras themselves are mounted on tripods and are the technical counterparts of the studio cameras but are considerably lighter in weight. The cameraman selects the scene through a finder on top of the box, while the focus is regulated from the control panel by means of a motor mounted on the side of the camera and connected to the lens housing. The camera is not so cumbersome as the newsreel movie machine. It is connected to the mobile unit through a special cable several hundred feet in length to permit the greatest possible range. The long cable also allows the camera to be taken indoors, although the mobile unit remains outside.

The second van, which houses the transmitter, operating on a frequency of 59 megacycles, is connected to the first with a coaxial cable and if necessary can be 500 feet away. On the roof is a collapsible antenna of the directional type. Signals pass from it in as nearly a straight line as possible to the main transmitter. There a special receiver, also using a directional antenna, feeds the big station. Because the transmitting equipment generates quite a bit of heat while in operation, the interior of the van is cooled by air drawn through filters at the rear and forced out through the front. A water cooling system is used on the tubes. The apparatus is also mounted in easily accessible racks.

This mobile unit has had one serious handicap, that of the amount of power needed for its functioning. This has been an important consideration, for a source of three-phase, 220-volt, 25-kilowatt power is not always readily available without a special installation. Another factor entering into the development of this type of remote pickup apparatus is the cost. One conservative estimate placed the amount involved in building the rolling transmitter at $125,000. This does not cover maintenance nor the money spent on improvements incorporated since the machines first made their appearance on a test basis in the fall of 1937.

To solve some of the problems and to overcome other obstacles encountered by this first American mobile television station, Radio Corporation of America engineers continued their research work, with the result that they developed what they describe as "lightweight portable television field pickup equipment," a "vest pocket" station, so to speak. Instead of two motor vans, the apparatus is installed in carrying cases much the same as in the remote-control field equipment often used in sound broadcasting. The cases weigh from 35 to 72 pounds, depending on the type of material they contain. An ordi-

nary station wagon or light delivery truck is large enough
to transport the complete station. Thus, weight has been
reduced to one-tenth, and what is just as important, the
cost is one-sixth that of the two-van transmitter—a little
more than $20,000.

Furthermore, the power problem has been solved. Instead of the 220-volt, three-phase requirement of the
older equipment, the new operates on 110-volt, single-
phase, alternating current. That makes almost any house
lighting circuit an adequate source of power supply.
Thus, when afield there would be scarcely any place
where current connections could not be made. Power
consumption is no more than 3.6 kilowatts compared to
the 25 kilowatts of the other. While designed to function
from regulation supply sources, it would be entirely practical to carry along a generator to make power, particularly in cases where the equipment is housed in a light
truck. This would make the station independent of outside facilities and would make telecasts possible in instances of power failure or where no power was readily
available. In that way there would be none of the handicaps that might otherwise be encountered in floods or
other occurrences when telecasting.

The operating frequency has been changed, too. It is
288 to 342 megacycles, about one meter. This channel does not encounter the same types of interference

experienced on other waves. That is, electrical disturbances and lightning static are not serious factors. Small but highly efficient antennae multiply the effective power of the transmitter several times.

One of the first jobs handled was an airplane assignment, in which the unit was taken up over New York City to televise scenes far below. Another was a demonstration test before members of the Federal Communications Commission in Washington. The setup incorporates part of a movie technique, the fading in and out of different camera angle views. One scene builds up on the screen as the other slowly dissolves.

After its public introduction in televising New York City from an airplane in a 45-minute transmission on March 6, 1940, the suitcase transmitter was given an important assignment at NBC. This consisted of paying visits to sound broadcast studios and televising network programs that could not be moved into the television studio.

The station is designed to handle one, two, or three cameras, one more than the "telemobiles." A one-camera assembly for relay over wire lines has but four carrying cases, each of which is about the size of an ordinary suitcase. The weight is roughly 275 pounds. In addition there are the camera and the connecting cables. Incorporation of a second camera adds four more cases, in-

creasing the weight to 548 pounds. The third camera requires but three more cases, the weight then going up to
862 pounds. The newly developed relay transmitter and
its power unit add only 250 pounds. So, with 1,000 feet
of cable to hook in the cameras and other connectors for
the various suitcase sections, the over-all weight is less
than 1,500 pounds for the complete three-camera field
station. The "telemobiles," handling only two cameras,
weigh nearly ten tons.

Besides all of the improvements with their flexibility
of operation, the engineers had in mind the televising of
night club floor shows, of dramatic productions and
musical shows on adequately lighted Broadway stages;
this is more easy of accomplishment with the new unit
than with the old.

To start with, RCA engineers put together two of
the new stations. One, using two cameras, went to the
Don Lee Broadcasting System at Los Angeles, where
Harry R. Lubcke has pioneered in Pacific Coast programming since 1933. He tells about it in another chapter, but
it should be recorded here that his portable equipment
was given its inauguration at the 1940 Rose Bowl Tournament of Roses Parade at Pasadena, sending the scenes to
the main station for transmission to receivers in and about
the land of the movies.

The second, also a two-camera station with provisions

for an additional camera, was assigned to NBC to augment its already elaborate equipment, while a third was arranged for the Columbia Broadcasting System to be adapted to a new type of mobile unit to be completely self-powered and to carry three new type and highly sensitive cameras.

Although the suitcase design was followed by the engineers in building this new equipment, they laid plans for its permanent installation in an automobile, but just one automobile this time, and not of the bus type. To quote from the engineering description: "The relative simplicity of the equipment and the lightness of each unit make the apparatus much more easy to handle and to transport than any previous equipment. . . . The apparatus is capable of producing high quality pictures comparable under good conditions to those produced by standard studio equipment."

It has not been the purpose of this section to go into questions as to whether television should be operated with this or that system; neither has it sought to discuss the detailed technical make-up of the mobile units, old or new. Suffice it to say that the system used to date has performed exceptionally well.

Little consideration has been given to progress made in England, because at the time this chapter was written, activities were in abeyance because of the war. However, it should be kept in mind that British television was publicly available for some time before the same step was taken in this country. They started regular programs a year or so ahead of us, and in that time did considerable outdoor televising. One of their milestones was radio picturization of the coronation parade of King George and Queen Elizabeth. Numerous other features were handled, including sports.

To the reader it might seem that there has been considerable concentration, possibly too much, on the New York scene. But that's where the activity described has taken place through the facilities of the National Broadcasting Company and its parent company, the Radio Corporation of America. However that may be, it is certain that the experience which has been gained in and about the country's biggest city will prove valuable to other television broadcasters of the outdoors when radio pictures become more uniformly distributed.

It has been extremely expensive, this bringing of scenes from afar into the living room. That is apparent from the cost of mobile unit No. 1, not taking into consideration the maintenance expense involved while afield, or the main transmitter.

However, the men who have operated NBC's outdoor televisor have gained wide experience in their first year's efforts. They have been doing a constantly improving job under the guidance of Burke Crotty, who in television compares to the director on a movie lot. In sunshine, in rain, in sleet, they have been out with their equipment sending through the air pictures of things as they happen. Whenever they go on a "job," particularly a first-timer, they often have to prepare weeks in advance to determine if conditions will be satisfactory for functioning of the cameras and for sending a signal that can be relayed to the main transmitter.

Crotty, an enthusiastically dynamic, sandy-haired young man, has had plenty of obstacles to overcome, for his directing job was the first of its kind in America. His basic experience was in the dissemination of photographs for publication purposes. This apparently proved a valuable background for developing the "newsreel sense." Close as he has been to the New York scene, there is no one who can better contribute firsthand comment. He says:

"Television's job in the new newsreel is now, and I believe will continue to be, a highly selective service. We must give a full or nearly full coverage of scheduled events, probably with a heavy emphasis on political events, sports, parades and similar features.

"The future is tightly bound up with the general American expansion of technical facilities. Hours of daily service must be lengthened and the television network must reach out to embrace the entire United States. We need have no fear that the engineers will not provide us with smaller and more sensitive cameras as well as lighter and more flexible relay transmitters. These things will surely come and the day will arrive when television will enable a viewer in Seattle to look in on some important happening, probably even in Paris or London.

"As for television's creation of the new newsreel, its incomparable quality of instantaneous transmission of things happening has given it a place secure in the scheme of things now and to come."

A chapter such as this would not be complete unless an attempt were made to peep into the future. It certainly is within the realm of possibility that during the next decade we may see a nation-inspiring event in almost any place in the land. All that such an accomplishment awaits, as Crotty points out, is an increase in the number of stations and the development of an interconnecting network. The engineers have demonstrated that this can be done in two ways, by ultra-short-wave radio and by the coaxial cable. Progress in their construction to give the necessary facilities is a slow process. The speed of development will depend upon public acceptance of television.

An event upon which the whole nation may look some day is the inauguration of a new president. One could go even further and look forward to the day when presidential campaigns may be conducted almost wholly by television. Radio already has demonstrated that a candidate need not be an ambitious traveler to make himself heard all over the country. Eyes of the air might eliminate the traveling altogether. With television fully established, the candidate could be seen almost anywhere. The one who is most telegenic might even be the top vote getter.

There is another event that gains annual nation-wide attention, the World's Series. It certainly would make ideal television material. Another is the New Year's Day Rose Bowl football game at Pasadena.

One could go on and on through a long list of possibilities from outdoors. And as television spreads over the land the subjects available for the out-of-the-studio cameras will become almost unlimited. Then the home will be as close to anywhere as the television receiver.

Despite the obstacles and the problems, financial, technical, and, in instances, physical, the new newsreel has got off to a good start. The pace it follows henceforth depends as much upon you, the reader, as it does on the men behind television. If you become as avid a looker as you have been a listener, the progress made will be outstanding.

The next chance you get, take a look-in on somebody's

receiver—it could be your own—and see for yourself what outdoor television is doing. It has its limitations. Nothing that man makes can be perfect. Nevertheless, it might surprise you by the great vista it is opening up—this, *the new newsreel.*

CHAPTER EIGHT

Television on the West Coast

BY HARRY R. LUBCKE
*Director of Television, The Don Lee
Broadcasting System, Los Angeles*

THE television activities of the Don Lee Broadcasting
System began in 1930. With a ten-year history of development behind it and being located in Los Angeles,
Don Lee holds a significant position. Since Hollywood is
a next-door neighbor, the film industry has gained its first
impression of television programming from the Thomas
Lee-owned station, W6XAO.

Actors, directors, and cameramen from the cinema capital have been quick to express interest and to offer co-
operation. Being adjacent to Hollywood, W6XAO is in
the midst of a treasury of talent which has proved of
great value and interest in working toward the ideal television program. Then, too, Hollywood is deeply concerned with the direction television takes. Realizing that
its future may be modified by television, Hollywood
stands by, waiting and watchful. Contrary to certain
schools of thought, we believe that television will supple-

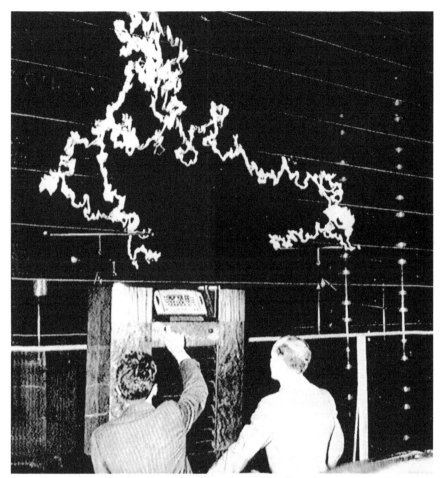

Frequency modulation receiver undergoing comparative listening tests while subjected to a million-volt lightning discharge. (COURTESY GENERAL ELECTRIC CO.)

Oscillograms demonstrating difference between amplitude modulation (*left*) and frequency modulation (*right*) reception. Relative noise level of each is indicated by the visible distortions. (COURTESY GENERAL ELECTRIC CO.)

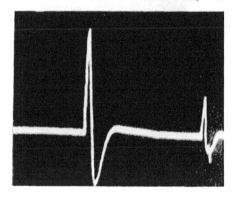

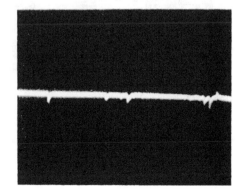

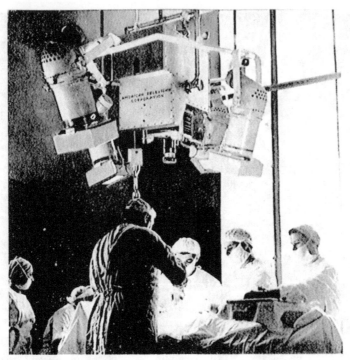

The television camera suspended over the operating table at the Israel Zion Hospital in Brooklyn, N. Y., enabled medical students in another part of the building to witness close-ups of delicate surgical operations. (COURTESY AMERICAN TELEVISION CORPORATION)

Large screen projection television installation in a theater. (COURTESY BAIRD TELEVISION CORPORATION)

ment rather than supplant the motion picture industry. It will no more eliminate the motion picture than the telephone eliminated the telegraph.

The Don Lee Television Station W6XAO is located in an eight-story limit-height building on the side of a gently rising hill, 305 feet above sea level. The television antenna is a unique paddle-wheel type, 200 feet higher. Station W6XAO has been on the air since December 23, 1931. Over 2,600 programs have been telecast to the estimated 400 receivers within a 30-mile radius of the station. More than 6,000 scheduled program hours of television have already been broadcast, which is believed to exceed the activity of any other television broadcaster in the nation.

Original technical research has played a large part in the activities in addition to program experimentation. One of the most interesting technical experiments occurred in May, 1932. The Don Lee staff wished to demonstrate the self-synchronizing characteristics of a receiver which they had developed. In order to test it under the most severe conditions, it was taken aloft in a tri-motored transport airplane. Images were received from the W6XAO transmitter while flying over the city of Los Angeles. It is believed that this was the first transmission of television images to an airplane and the first observance of television images while traveling at airplane speed.

As a result of this experiment, many interesting facts were discovered concerning ultra-high-frequency airplane television reception, the most significant of which was the presence of ground and obstacle reflections.

The active program service of the Don Lee television station began on March 10, 1933. On that date an earthquake shook Los Angeles and damaged the near-by towns of Long Beach and Compton. Immediately, a film organization was contacted which went to and photographed the scenes of disaster. These films were then broadcast from our station. Night scenes of the distressed areas were shown first, and then day scenes. Later, two full reels were telecast which gave a comprehensive picture of the whole damaged area. These broadcasts were received in several stores and homes throughout the city. Two stores held public demonstrations of the event. Since the public was not admitted to the stricken area for some two weeks, many saw television images of the damage before they were permitted to visit and inspect the actual scenes. This is believed to be the first television transmission of a major disaster, whether from film or otherwise.

From this rather striking inauguration, the Don Lee System has been working constantly toward perfecting the "format" of a good television program. Mr. Thomas Lee, President of the Don Lee Broadcasting System, believes that the television program of the future will be neither strictly drama nor strictly film as those forms are employed in the entertainment world today. He foresees a program content that will offer "contrast" in successive sections with the view of maintaining interest in the television screen.

One of the features of the Don Lee television schedule is what we term "living-room education." Experimental programs of this type have already met with enthusiastic response. Our staff is working on the theory that Americans at home will enjoy specially and intimately presented educational programs that would not draw them to schoolrooms or motion picture theaters. In line with this, live subject demonstrations have been presented by authorities from the University of Southern California, Los Angeles schools, art centers, and other organizations. Documentary films have proved an attractive feature in this type of public service. Demonstrations have presented such subjects as the lie detector, making pottery before the camera, first aid, golf strokes, origin of the alphabet, charcoal drawing, and facial cast making. Domestic demonstrations such as cake frosting and pastry making have also proved popular. It is Mr. Lee's conviction that programs like these will continue the education of the American adult in an attractive manner and avoid the unfortunate cessation of the educational process on graduation from formal school.

Classroom television has already been attempted. One program was transmitted to four schools. It was a combination of live talent and film presenting the story of the trans-Pacific Clipper airplanes.

Other features planned to give variety and public service center around our recently acquired "suitcase" type mobile unit pickup equipment. One remote pickup a week is usually found on our program schedule. In addition, the eventual goal is a nightly spot newsreel service based on topics of local interest. This would be telecast in the early evening when the family is at home and at leisure to receive the events of the day. Immediate plans include mobile pickups of the filming of a motion picture with first-hand views of the many intricate and fascinating angles of motion picture technique, as well as daylight transmission of outdoor events.

With mobile unit equipment, we have already successfully televised the Pasadena Tournament of Roses Parade on New Year's Day, 1940. Although rain fell, hundreds of Los Angeles television receiver owners saw the parade in comfort in their homes. The portable equipment "beamed" the images nine miles from the scene to the main station W6XAO, whence they were broadcast to the homes over the usual wave length. The telecast was two hours in length, the major portion being run simultaneously with the nation-wide radio broadcast over the Mutual Network.

Subsequent pickups have included televising the Easter sunrise service from the Hollywood Bowl, baseball games of the Coast League at the Hollywood home grounds, wrestling and boxing matches from the Hollywood Legion Stadium, and a rodeo from the Los Angeles Coliseum.

Our nearness to Hollywood affords unique opportunities for mobile unit material. The brilliant premieres of Hollywood offer a never-ending source of interest to our viewers, and the mobile unit will bring the colorful evenings thronged with famous personalities into the living rooms of our audience.

Several radio programs have been televised while they were being broadcast by sound over the nation-wide Mutual Network. This is part of the process of making the transition from sound to sight-sound broadcasting. These have included songs by Maxine Gray, by Betty Jane Rhodes, and by the Sons of the Pioneers.

To approach the perfect television program format, every type of live production has been given, from a two-act costume drama to a demonstration of how to handle live rattlesnakes. *Macbeth*, with Fritz Leiber in the title role, was one of the most successful television program items. More of these are planned. Our Teletheater Guild Unit has produced weekly experimental dramatic production programs. These have included *Alice in Wonderland*, in three episodes, and a novel presentation of a group recital of blank verse with synchronized gestures and music.

Other unusual programs have included plays produced by Max Reinhardt and his workshop, now located in Hollywood. The Pasadena Playhouse has also presented experimental plays. Our program usually includes four dramatic sketches per week.

One of the best-liked entertainment features of W6XAO is the television serial *Vine Street*. This has developed into a significant experiment in adapting the serial story to television. Many problems of television technique have been met and solved during the semi-weekly fifteen-minute episodes of *Vine Street*. In the belief that a brief history of how this television serial was organized will interest the reader, we here give a few details.

Vine Street is a comedy drama of Hollywood life written by Maurice Anthony and Wilfred Pettitt, and starring Shirley Thomas and John Barkeley. The serial has many followers among the Los Angeles viewers who have followed the principals from poverty to riches, from park bench to movie studio, from San Fernando to Hawaii.

When the idea of a television serial was conceived, the first step was to organize a separate "dramatic unit" for the purpose. Actors were interviewed and the unit assembled. A type of production was selected which was best suited to the talents of the actors we had chosen and also fitted to the station's program needs. The writers and the cast were then assembled. Two weeks was spent in the preparation of a sample script and conferences were held with the television production department. A dress rehearsal was then held and the necessary modifications were made. Then the unit went into production.

Episodes of *Vine Street* are presented on Tuesday and Friday evenings. They are of fifteen minutes' duration. At least two, and often three, episodes are prepared in advance. The content and action of the forthcoming episodes are determined by a story conference of cast and writers. The general theme of the serial is considered and the ideas, dialogue, and scenes are evolved. The writing is done subsequently by Wilfred Pettitt, chief writer for the unit. The script is complete with respect to dialogue, and camera shots and special effects are indicated. At times, sketches of the scenes as they are to be taken by the camera are included. In writing, cognizance is taken of the fact that large and elaborate sets are beyond the present scope of television economically, and that physically impossible actions must not be imposed on the cast.

The characters learn one episode at a time. The principals spend several hours two days preceding the broadcast in memorizing their lines, and a short period going through the dialogue together and establishing gestures and action to fit.

The day before the broadcast an hour is spent in refining the recitation of dialogue. That evening, the complete staff rehearsal is held. This includes the camera and lighting crew, the sound supervisor, and the sound effects man, under the direction of the television producer.

The first activity at this meeting is the distribution of the scripts to all concerned, and a résumé of the important camera shots and actions, special lighting, and sound effects, as recommended by the cast. These recommendations are modified by any of the several staff members as required. Necessary changes are agreed upon and the important aspects of the episode tested under television transmission conditions. Following this a dress rehearsal is held to familiarize and correlate the operative production with the action, and it is in this portion of the production that the greater part of what appears on the television screen is formulated. The appearance of the properties is checked on the television monitor screen and the lighting arranged or the properties modified, until a satisfactory delineation is secured. The lighting and position of the cast and the camera angles are also determined. Microphone positions are established and sound effects tested.

On the night of the broadcast, the principals arrive two hours ahead of the scheduled time for the episode. Last-minute modifications are made with the operating staff and new recommendations from the staff are received.

The production department, in co-operation with the stage manager, ascertains how the properties and scenery should be handled with respect to the rest of the television program of the evening. Usually *Vine Street* is the last act on the program. The scenery and properties are often placed on a set prior to the broadcast, and this set is not used for other acts. The cast is then made up.

A visual and aural introduction is provided for the episode by means of a theme, a miniature stage, and an appropriate introductory paragraph, which is prepared by the writer of the script. Concurrently, an appropriate World Broadcasting transcribed theme, "Sophisticated Lady," plays on the aural channel. In motion picture title fashion, a miniature stage starts the performance by the raising of the main curtain, the draping of a side curtain, and the retraction of side wings. This reveals a sign reading, "*Vine Street* by W. H. Pettitt." The side wings are then moved to obscure the sign, which is immediately replaced by a second sign reading, "Starring Shirley Thomas as Sandra Bush." In the same manner her photograph is next displayed, then a sign reading, "and John Barkeley as Michael Roberts," which is followed by a photograph of Mr. Barkeley. Simultaneously with the visual action, an offstage announcer ties the forthcoming episode to the previous action and introduces the episode. A second camera then takes the scene, the action starts, and carries on to conclusion.

We have been making daily tests over a period of years to ascertain the suitability of films available for television broadcasting. The results of this study were presented before the American Society of Cinematographers on August 30, 1937. The members of this organization photograph over 90 per cent of the motion pictures made in Hollywood.

At this conference, seven rules for television cinematography were presented. These are quoted below in the belief that the results of this research may be of value to those interested in television.

Rule I: Do not violate the usual rules of photography. Illumination, composition, contrast, and exposure are to be used as required for clear pictorial definition. In motion picture photography, extremes in lighting and other factors are employed for dramatic effect. Dark, low-key lighting is used to produce a depressing audience reaction to tragic sequences. Such practices may be employed to a limited degree in television technique, but they must be modified or the result at the receiver will be unsatisfactory.

Rule II: Carry detail in the half-tones. The object of principal interest must be portrayed in this manner. For instance, the outline of a man in a black tuxedo is lost against a black background drape.

Rule III: Achieve "checkerboard contrast." This is a form of composition based on the knowledge that the whole field of view is broken into alternate dark and light areas. The name originated because of the clarity with which a checkerboard, held on the laps of two convalescent soldiers, was reproduced in a scene from our early researches. It is not necessary that the various areas be of the same size or symmetrically distributed.

Rule IV: Keep the over-all contrast range small. This rule is frequently violated in taking personage shots on shipboard. Here a dark figure is often secured with a "clear celluloid" background. Such extremes encounter overload points in the several units of the television chain from pickup device to receiver screen.

Rule V: Maintain action. Television as a medium requires a faster tempo than the screen. Attention may be diverted from the receiver by household annoyances during periods of inactivity. Consequently, the plot and characters must carry the story forward at an interesting pace.

When inanimate objects are to be shown, motion of the camera can fulfill the rule. "Panning" is effective and desirable in scenic exteriors. Panning, "zooming," change of camera angle, and traveling shots all supply variety in interiors and also may be employed in exteriors.

Rule VI: Supply medium or light density prints with black frame lines. Dark prints are definitely inferior to lighter prints from the same negative.

Rule VII: Employ lap dissolves, quick fades, or change instantaneously from scene to scene. Long fade-outs may give the audience the momentary impression that something has gone wrong with their television receiver.

Public demonstrations of television have been held on the West Coast for a long time. During the year 1936 the Don Lee Broadcasting System undertook extensive demonstrations of television on the main floor of the Don Lee Building in Los Angeles. There was no admission charge. During the month of June, 1936, about 5,000 people witnessed these demonstrations. The visitors were a typical cross section of the American public. Many were attracted by signs in the windows of the Don Lee Building inviting the public to our demonstrations. Others came in response to notices in local newspaper columns. Many of these people were seeing television for the first time. Their reactions were largely favorable and many requested information as to where receivers could be purchased. Although receivers were not for sale, and have never been sold by our organization, many skilled persons constructed their own television receivers from plans made available without charge. This group still forms an appreciable portion of our television audience, and from its ranks have been recruited television service engineers who now install and service commercial television receivers.

In April, 1937, this demonstration service was moved to the California Institute of Technology, in Pasadena, ten miles away from W6XAO. Here demonstrations were held for the scientists of this institution, and for 2,500 persons of the interested public.

Continuing this service, a receiver was installed in a large private residence about four miles from the transmitter. Here, apart from the living room, a "television room" was created in which the receiver stood. Two rows of chairs faced this in the manner of a miniature theater. Demonstrations were held on many evenings with as many as eight separate showings in a single evening. Admission to the demonstrations was by tickets, which were obtainable without charge. In this intimate setting, more than 2,500 people made their first acquaintance with television over a period of three years. Many individuals representing the general public, motion picture and radio stars and executives, as well as groups from such organizations as the Academy of Motion Picture Arts and Sciences, and the Institute of Radio Engineers, attended these evenings of television entertainment.

In addition to these presentations of television to the West Coast public, a nonprofit organization known as the Hollywood Television Society has given weekly public demonstrations at the Don Lee W6XAO television transmissions for the last two years.

Now, leading department stores and some twenty-five dealers, wholesalers, and manufacturers hold demonstrations on nearly every W6XAO television transmission. Often fifty persons will be gathered before a single receiver. On February 8, 1940, a wrestling bout over W6XAO drew such public interest before a radio store in Long Beach (twenty miles from the station), where a receiver was displaying images in the window, that the police were required to control the crowd and allow traffic to proceed.

Extensive plans are being made for future television operations on an ever-increasing scale. Most important of these is the construction of a scientifically planned television station on Mt. Lee, which is two and a half miles from the heart of Hollywood. Mt. Lee was selected after carefully considering the best site for hilltop television transmission to which power, water, and telephone facilities could be brought and which the city planning commission would approve. The nearest house is five-eighths of a mile airline from Mt. Lee and the sides of the mountain are so steep that it will be impossible to build houses closer than a quarter of a mile from the peak. Yet 43,452 people reside within three miles of this mountain, and 596,002 people within eight miles. This is an ideal situation, a high natural eminence of 1,700 feet, located near the center of a population of a million. Within fifty miles the population reaches two million. It is expected that an area of fifty or sixty miles will be dependably covered. Mt. Lee is one and a half times the height of the Empire State Building in New York City.

In designing the Mt. Lee Studios visits were made to major motion picture studios and their technique studied. In some respects their lead can be followed; in others the differing requirements of television production indicate other procedure. In common with motion pictures, provision must be made for sound deadening and adequate ventilation. In television, however, immense stages are not required nor even desired since the inescapable echoes arising in such stages are a hindrance to good broadcasting. The design of radio studios should not be followed since these are often too small dimensionally and too "live" acoustically for television use.

The plans of the first Mt. Lee structure call for a two-floor building housing two stages, dressing rooms, film room, control booths, transmitter room, offices, laboratory, and scenery and properties storage rooms.

The building will be located at the center of the extended flat top of Mt. Lee. The road is already being improved to first-class condition, and water, electricity, and telephone facilities will be extended. Most of the preliminary investigations were made during the spring and summer of 1939. Plans for the building have been tentatively approved and engineering calculations on the roof trusses and columns completed. With the completion of this elaborate television project, the West Coast will have one of the best-equipped stations in the country.

An application has been made to the Federal Communications Commission for a permit to construct a new television station in San Francisco. This would make possible the first West Coast network.

Current broadcasts from W6XAO are for a minimum of ten and one-half hours per week. Afternoon and evening broadcasts are made on Tuesday, Thursday, and Saturday, totaling two and one-half hours per day. Monday, Wednesday, and Friday, there are one-hour evening broadcasts. Each program includes live and film subjects; the first three days of the week being devoted predominantly to "live" subjects, two days to "film," and one or

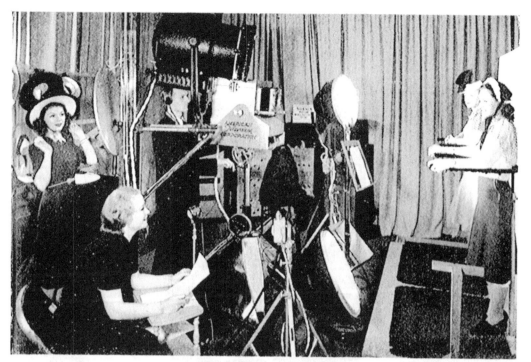

One of the first commercial adaptations of television is its use in showing merchandise, modeled and displayed on receivers throughout large department stores for the convenience of shoppers. (COURTESY AMERICAN TELEVISION CORPORATION)

Television broadcasting station W2XB erected by the General Electric Company atop 1,500 foot hill in Helderberg Mountains adjacent to Schenectady, Albany, and Troy, New York, to cover area of 500,000 people. Many of NBC's New York programs will be rebroadcast from this transmitter. (COURTESY GENERAL ELECTRIC CO.)

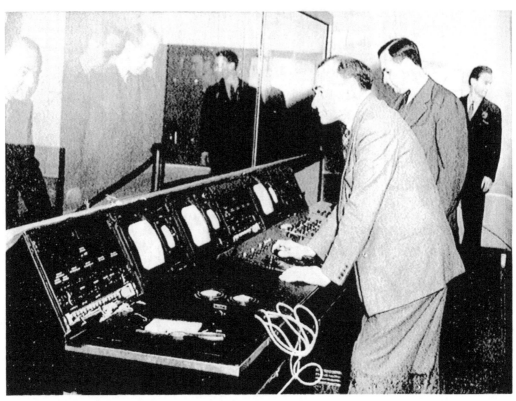

Gilbert Seldes, Director of Television Programs for the Columbia Broadcasting System, at the control panel of the CBS television studio in the Grand Central Terminal Building, New York. (COURTESY CBS)

The CBS television antenna atop the Chrysler Building. (COURTESY CBS)

Special television camera designed by CBS engineers. (COURTESY CBS)

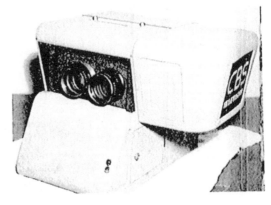

more to outside pickup events. Two studio cameras and one film camera are used at present.

A recent program was viewed by Bernard H. Linden, inspector in charge of the southwestern district of the United States and special representative of the Federal Communications Commission to visit Station W6XAO in a nation-wide inspection of television by the Commission to determine the state of the medium. Inspector Linden made this statement for the press: "I was exceedingly well impressed with the performance of the television show of W6XAO both at the sending and receiving ends. I marveled at the lack of interference with reception, and as long as I was at the station and at the home of Thomas S. Lee, 7 miles away, the images were very clear and realistic and I enjoyed the performance as well as the skilled technique demonstrated by the staff in handling the equipment."

With its pioneering work in both technical and program research, its efforts to make ever larger groups of people television-conscious, and its project of the scientifically planned Mt. Lee television station, the West Coast takes its place as a focus of American television activities.

CHAPTER NINE

Facsimile and Frequency Modulation

BY J. R. POPPELE
Chief Engineer, Radio Station WOR

TWO new developments that are, with television, rapidly transforming our existing methods of communication are facsimile and frequency modulation. While they are separate services, all three systems bear a relation to each other. Their futures are intertwined. New to the public, they hold in common the promise of vastly changing the entire communication system, from which they are as different as modern radio is from the early crystal sets.

One way of describing facsimile is to say that radio has learned to write. Frequency modulation, the second development mentioned, is a new method of transmission and has been referred to as "staticless radio"; this is a good shorthand definition, although it offers a greater improvement than the mere elimination of static. It may be the key to the unification of television, facsimile, and

Facsimile and Frequency Modulation

sound broadcasting. It offers the possibility of the three services, sound, sight, and printed material, being transmitted simultaneously without marring their separate smooth reception. With the combination of these three, a stirring telecast of a news event may appear in your home accompanied by such material as the printed story, photographs, maps, etc., coming out of your set at the same time that you are watching and listening to the event. This is no inventor's dream: it will be possible with the further development of technical progress in the field. We are living on the threshold of tremendous developments and wide changes in the system of air wave communication. The actual changes will, of course, be gradual. In the case of frequency modulation, the early receivers will receive both frequency modulation and the present broadcast bands. The first stations using this new system will simultaneously broadcast by means of frequency modulation the same programs as are regularly transmitted on the present radio channels. Gradually separate program services will be developed; the lapse of time between the change from the existing system to the new one will of course depend upon the speed of set sales and upon public acceptance.

We Present Television

Facsimile

Facsimile, the service that transmits a miniature newspaper through the air waves into your home, can be received by means of an attachment to your present radio receiver. Already many facsimile stations as well as receivers are in operation, and newspapers are experimenting with this system as an auxiliary means of getting news out fast to the home in permanent printed form. Radio "readers" receive the latest in world events, pictures, sports, drawings, and educational material through this service.

Radio Station WOR, one of the most advanced of the facsimile-broadcasters, has entered the field with two of the foremost facsimile systems. The first, devised by W. G. H. Finch, employs an electrolysis process with specially treated paper, four inches wide. The second, a development of the Radio Corporation of America, prints a page eight inches wide by the use of a stylus and carbon paper.

Facsimile was first suggested by Alexander Bain, an English physicist, back in 1842. His work is still the basis for the present-day systems.

Facsimile is based on principles similar to television. The copy to be transmitted, which may be straight reading matter, photographs, or line drawings, is inserted in a scanning device. Then an electric bulb sends a pin-prick of light shuttling back and forth across the copy, much as the human eye sweeps back and forth across a printed page. The differences between light and dark are minutely noted by a photoelectric cell to which the light is reflected, and these variations are translated into electrical energy.

From this point on, the process is similar to regular sound transmission, the output of the scanner being treated in the same fashion as that of an ordinary microphone. Eventually this signal modulates the radio transmitter carrier wave and is broadcast. Any receiver which tunes to the wave length used is suitable to intercept the facsimile signals. Here again the process does not differ from sound transmission, except that in place of the loud-speaker a special facsimile printer is used to change electrical energy back into the original copy. This is accomplished by a stylus sweeping synchronously with the transmitting scanner across the paper, with a black impulse depressing the stylus, and a white one lifting it. About a hundred strokes are used to build up one inch of facsimile copy. The result is a small printed record which you may read, clip, and keep for future reference.

Facsimile can be broadcast by the transmitters of the present radio stations during such time as they are not on the air with sound programs, usually in the early morning hours. When used on the long wave lengths that these stations use, the signals are not subject to the distance limitations that television encounters on the high frequencies. Facsimile signals from Station WOR have been received in a yacht in the Gulf of Mexico, and in homes in California. Facsimile is also broadcast by stations on the high frequencies.

Facsimile has been used as a method of transmitting information to the police in radio cars and sending maps and weather data to airplanes in flight. It is used in press services on telephone or radio channels to transmit news as well as black and white and color illustrations. In the classroom, facsimile receivers have cut mimeograph stencils and the copies have been distributed to students for the latest information on many subjects. President Roosevelt when on board ship has received his latest news by this service. These are but a few of the many existing uses of facsimile, which has unlimited possibilities for many other uses in the future.

Inexpensive facsimile receivers are now available for the home, and it is expected that this new service will soon be in universal use.

Facsimile and Frequency Modulation

Frequency Modulation

The system that is used to give us our present radio programs is known as amplitude modulation—called AM, for convenience. The new system is called frequency modulation, or FM. You have to hear frequency modulation to realize the startling changes it makes in sound broadcasting. Musical notes come into your home clear, separate, and staticless. The illusion of having a fine musician in your home playing a superb instrument is uncannily real. The announcer whispers, and you start as if someone had spoken at your elbow. You can hear the striking of a match in the studio, and your radio does not crackle as if a firecracker were set off inside it. It merely records the sound with complete realism. The intake of breath as a smoker lights his cigarette is distinctly audible. How is this accomplished?

Frequency modulation makes possible the transmission of all the audible sound frequencies without static or background noise. The present radio transmitters use high power to break through this static, while frequency modulation stations, using little power, sidestep it almost complctcly, thus making possible the reception of even complete silence.

To clarify the technical differences between the two

methods of broadcasting, we may use the analogy of a searchlight operated with and without a shutter. A regular amplitude-modulated station upon your dial at, say, 1,000 kilocycles stays permanently in that wave channel and is thus beamed steadily at the listener's home. The intensity of the light would be varied by the shutter in accordance with the words and music being broadcast. Frequency modulation has an entirely different principle. The intensity of the light would stay constant while the beam would swing continually back and forth, depending upon the voices and music to be transmitted. In broadcasting FM, a much wider carrier wave results. Because there is considerably more room for this variation down in the wide-open spaces of the ultra-short waves, FM broadcasting of the future will continue to stay down there. A frequency-modulated station requires a channel or radio roadway about twenty times wider than the regular ten-kilocycle broadcast band channel for best results. Paradoxically enough, this does not really waste space on the air but makes more economical use of it, by allowing more stations in one area in one wave length.

This latter advantage is one of the most fascinating possibilities; it is the solution to the present-day problem of the overcrowded broadcast band. For the past twenty years radio has been growing and expanding until

Facsimile and Frequency Modulation

now there are so many stations jostling each other in the ether that the situation has become acute. Because of the great number of radio stations already in existence, many small communities eager for local radio service cannot have it, owing to this overcrowding. Clashes of smaller stations with powerful ones cause cross-talk and static, so much so that many stations are required to sign off at an early hour so that other stations may operate without any disturbance to mar their programs. In many localities intermittent noises mar radio receptions. Oil burners, elevator motors, dial telephones, X rays, diathermy machines, and many other devices all contribute their buzzes, clicks, scratchings, and cracklings to the program you would like to hear clearly. Those who live in the country more than fifty miles from a powerful station experience great static disturbances that really ruin enjoyment of radio programs. Static crackles through with annoying regularity. With frequency modulation, these things simply cannot happen. Reception is always smooth. The system is not susceptible to these static-causing interferences.

To give frequency modulation a really severe test and to introduce it vividly to the public, the General Electric Company conducted a remarkable experiment at the World's Fair. A million-volt continuing arc of "man-made lightning" was created only a few feet away from a frequency modulation set while it was bringing in a program! A set was used that could be tuned to either amplitude modulation or frequency modulation. With amplitude modulation the set's loud-speaker translated the visible fireworks display into a crashing roar in which the program music was completely lost. The set was tuned to a station broadcasting the same program by frequency modulation—and the music emerged from the loud-speaker with almost perfect clarity and fidelity. In the background was a static buzz, too slight to mar the enjoyment of the program, audible principally to the ears which were straining to catch distortions; this was the only aural evidence of a million volts of electricity which was dancing and sputtering a few feet away. This feat can be fully appreciated when one thinks of the havoc created in a radio program by an approaching thunderstorm on a summer's night.

Then a second test was made by employing an oscillograph, the electronic measuring device which produces in effect a constantly changing picture of wave forms, changing electrical energy into visible images on the end of a picture tube in the same way that television images are produced. The images can then be photographed. By this method, signals were tested under static-producing conditions with the AM and FM systems. Barely noticeable on the oscillogram obtained from the FM set,

Facsimile and Frequency Modulation

the noise painted its distortions in pointed tracings on the AM oscillogram, proof for the eye in black and white of the presence of static in one case and its virtual elimination in the other. The results of these interesting tests are visible in the accompanying illustrations.

The revolution brought about by introducing this system affects every branch of the radio industry, but what is the connection between television and frequency modulation?

First of all, it is pertinent to static elimination in television, for television is subject to the same static that bothers you on your present radio. But in television this static is visual. It manifests itself in blurring, picture jumping, and actual complete fade-out at its worst. By introducing FM into the television system, the visual static could be as effectively eliminated as audible static is on radio by this method.

Aside from its possible future use for general television reception, FM will probably be used for the radio relay system to carry television programs from one city to another.

FM's strongest proponent and discoverer, Major Edwin H. Armstrong, believes that television transmission through frequency modulation is entirely possible and that it is simply a matter of time until the system is worked out through research and experiment.

We Present Television

No coverage of frequency modulation would be complete without a word about Major Armstrong himself. He is professor of electrical engineering at Columbia University and ranks as one of the most important inventors in the history of radio. Frequency modulation is his fourth major discovery, the others including the regenerative "feed-back" circuit which brought radio out of the crystal set era; the superheterodyne circuit; and the superregenerative circuit, widely used for ultra-high-frequency receivers. His story is that of a man who believed in his theories when the authorities were incredulous, and he has fought them through until now the radio industry must admit that FM is one of the greatest broadcasting advances made since the advent of the vacuum tube. There are now more than a hundred active and prospective FM broadcasters, the very lively nucleus of a system that promises to change broadcasting history.

Here then we come upon what may be the link between all the air wave services. By the future application of a very broad frequency modulation carrier wave, it may be possible to transmit and receive simultaneously sound programs and facsimile, or television and facsimile. This would be possible with the existing system, but only with complex apparatus.

We have already mentioned the possibilities of instantaneous transmission of more than one broadcast service through the use of FM. To make this concrete to you, let us imagine a cooking demonstration coming into your home by television. Expertly the chef mixes before your eyes a new dish, perhaps a salad that is an inspiration for a warm summer's day. He shows you the ingredients and the utensils he uses in preparation. You watch the skillful blending under his hands and listen to his anecdotes of cooking experiences as he mixes the ingredients. As he shows and tells you the story, your facsimile receiver delivers the recipe to you, printing it before your eyes as the chef mentions the ingredients and tells you the proper proportions of each. You see how the thing is done, and after the telecast is over you simply clip the recipe from your facsimile receiver and put it in your kitchen file of recipes. This is, of course, a very simple application of these wonderful principles of the new kinds of communication that we shall soon be experiencing in our homes.

The educational possibilities of these new methods of communication are so wonderfully extensive that one scarcely knows how to begin to catalogue them. We can certainly predict that schools will find them invaluable. They will enjoy a combination of visual instruction via television and simultaneous printed instruction via facsimile which will furnish a permanent record for study and review. This should be a complete revitalizer of educational methods. There is a fresh, startling quality about these air wave miracles that cannot help but stimulate educators, newscasters, program makers, and entertainment planners to new and more vivid forms.

All the air wave industries are completely surrounded with question marks at present. This is a period of transition and experiment. There are economic problems as well as engineering difficulties to be worked out before you can sit casually before new receivers that afford all the facilities we have described and enjoy more wonderful programs than you had ever imagined possible.

The intention here has been to indicate the trends that appear to be pointing to an astounding unification of our air wave systems of communication. One can safely say that the last twenty years have been an exploitation largely of the "long waves" in the narrow bands. We appear to have gone as far as we can in that direction both as to networks and as to sound clarity and elimination of static. Now the ultra-short waves are forcing themselves upon our attention as an almost unexplored continent of the air. We do not know how far these explorations will take us, but we are embarking upon them.

We believe that within the span of our own lives we shall see the intricate technical and economic problems

Facsimile and Frequency Modulation

solved so that one air wave industry will complement another. Perhaps this brief summary of the interlocking possibilities of air wave communications will stimulate you to an interest in the subject that will hasten the process of unification, since the active interest of the public is the final force that brings the engineers' experiments out of the laboratory and into the home.

CHAPTER TEN

The Finance Problem

BY BENN HALL
Editorial Staff, Radio Daily

WHILE scientists are solving most of the technical problems facing television and are rapidly developing a new and individual technique for visual entertainment, those concerned with television's business aspects are thoughtful about its future. Their problems can be conveniently divided into two closely related questions: Who's going to pay for television entertainment, and what will cause people to buy television sets? A new industry which may be a potent economic stimulant will flourish when the right answers to these questions are found.

A hint as to the possible answers may be suggested by a study of television activity in England up till the time of the war, when television was blacked out. This study of English television may well be supplemented by an investigation of American television in its first year.

English purchasers of television sets, as in America,

were painfully infrequent at first. But as better entertainment, including sporting events, was telecast, and as sets were improved or reduced in price, the English audience increased. Regular television schedules were started in England in the summer of 1936. About 1,350 sets were sold the first year. The second year saw some 6,500 sets leave dealers for private homes. Better shows and sets whetted the public appetite. Before the present war started, it was estimated that 500 sets were being sold each week and that 20,000 sets were in use. This means that after several years of government-supported telecasts, 20,000 sets had been sold and that the television industry was just starting to grow.

In America about 1,000 sets were sold within six months after television's introduction. This corresponds quite closely to early English sales. Many potential American and presumably many potential English customers were interested in the sets but delayed buying, hoping for reductions and better programs. Early sales were probably made to those people, particularly the ones with good incomes, who desired to be "first" with the latest invention.

Some English television observers declare that the relative success in telecasting popular events, such as the famed Derby telecast, aroused tremendous interest which resulted in increased sales. Television is still too new in

America to point out any one type of program and state authoritatively that it has helped sales, but the great popularity of certain sports broadcasts indicates that similar events here should spur the same interest.

While English television faced many difficulties, it operated on an entirely different financial basis from American television. Our television activities are supported entirely by private capital, while English television programs, like English radio programs, were financed by the government with taxes collected from set owners. No advertising was or is permitted. The introduction of television in England was, therefore, largely a governmental problem, although naturally private set manufacturers were concerned.

Here the problem differs fundamentally from the English experience with television. Our broadcasting business is a private industry, operating under government supervision. The government does not spend a cent for commercial radio, except for the upkeep of the Federal Communications Commission which regulates it. Broadcasting is supported by advertisers who buy "time" to present shows to promote the sale of their goods. It is their expenditure and the public's demand for their shows which make the American broadcasting system possible. Even sustaining educational, symphonic, and other programs are dependent on commercial shows, since without them station owners would have no funds with which to present shows of a noncommercial nature.

Television will eventually be likewise supported by advertising. There are several possible alternatives, but at present there appears to be little possibility of any of these methods being practiced. One such method is government-owned-and-operated stations, supported directly or indirectly by taxes. In view of the increasing anxiety over taxation, it is unlikely that any administration would seek to impose a tax on a hitherto free form of entertainment. Listeners realize that advertising makes possible shows costing up to $30,000 an hour for talent alone. Any measure to substitute taxed radio for free commercial radio would undoubtedly meet with immediate resentment—if that is not too mild a term—from the majority of voters.

Another alternative is a rental system. Just as the telephone company and other utilities rent a service to subscribers who usually pay a monthly fee, it has been suggested that television might be developed into a subscription-type service. Users would pay a regular monthly fee which would entitle them to a guaranteed amount of entertainment.

There are several objections to this proposal. Again the public would resent paying for such a service after being "educated" to expect free broadcast service. It is exceedingly doubtful if the average fan would "warm" to this method. Without mass support, acceptable programs could scarcely be presented at reasonable charges.

Furthermore, a single monopoly service would probably lack the necessary incentive offered by competition as it exists in present-day advertising. Dealing in television programs or in any other form of entertainment would be totally foreign to the usual public utility operation such as telephone or gas or electric service, where reliability and costs are the consumer's chief interests. Showmanship is, at best, highly intangible. It cannot be developed to the best advantage in a private monopoly without competition, or on a "civil service" basis in which a government bureau alone would decide what the public wants.

The surprising growth of privately owned radio stations on the Continent which, before the war, transmitted sponsored shows in English to British listeners, indicated that the government-operated stations by no means satisfied all English set owners. One of these stations charged higher hour rates than any station in the world, indicating a vast audience. It is of more than passing interest that many recorded American radio programs were popular on these stations.

At one time there was talk of a United States government subsidy to help in introducing television, but few experienced radio executives took these reports seriously. Such a move, it was feared, might pave the way for eventual government ownership of the entire broadcasting structure, and the plan never received general support. Because of the public acceptance of privately operated radio stations, it appears most unlikely that any other method of television practice would meet with approval.

There is little doubt that the American public will pay for television entertainment—indirectly, of course—through the purchase of sets and through supporting television advertisers. And now that plans have been made for drastically reducing set prices, large sales increases are expected. But since between $15,000,000 and $25,000,000 have been invested in television research, it is to be expected that further investments will be made to insure a return on the original investment. The public's hesitancy in purchasing sets taught television broadcasters one vital fact—that sets must be available to everyone, and that entertainment must be provided which will make a television set almost indispensable in many homes. The solution for these self-evident facts is being found.

A general picture of the present television setup may furnish a clearer picture of the entire industry and its aim.

We Present Television

The Radio Corporation of America has probably made the largest investment. RCA is interested in (1) selling receiving sets; (2) selling transmitting equipment to other stations; and (3) selling time, through its affiliate, the National Broadcasting Company, to sponsors. The General Electric Company has similar aims. The Philco Radio and Television Corporation will probably be primarily interested in the sale of receivers, although it operates an experimental station in Philadelphia. The Columbia Broadcasting System is primarily interested in the sale of time, as is the Don Lee Broadcasting System. Farnsworth Radio and Television, Inc., probably hopes to realize its greatest profits from the sale of receivers, while the Zenith Radio Corporation is likewise looking to future receiving set sales. The Allen B. DuMont Laboratories, Inc., will probably realize considerable revenue from set sales. The position of this company is particularly interesting because of important patents which it holds and because it is affiliated with Paramount Pictures, Inc. There are other companies interested in various phases of television but obviously those with the largest investments are most eager to see it gain public acceptance. Entertainment is the second phase of the television problem and will be another costly item. It may not, however, be so expensive as some critics have predicted. There are several reasons for this. One revolves around the suitable number of hours of daily television entertainment. Another factor is connected with the actual costs of entertainment programs. Television production staffs are literally working day and night to develop attractive and inexpensive programs; however, no entertainment medium has ever been developed overnight but is the result of collective endeavor, both artistic and scientific.

There have been far too many misconceptions about television entertainment, particularly when even those who have been experimenting for years will concede that there is still much to be learned. Figures have been bandied about freely, often based on the assumption that it will be necessary for television stations to provide continuous entertainment, just as radio stations furnish programs sixteen hours a day. It is doubtful, however, if the human eye could endure the constant strain of looking at small images, no matter how clear they are. It is also doubtful if any one person would care to concentrate on television more than a few hours daily. While radio often forms the background entertainment for conversation or cardplaying at home, or offers a diversion to the woman doing her housework, television usually demands full attention. It is possible, however, that with the increased telecasting of regular sound programs to give the added sight, most programs may be eventually combined with both sight and sound. Sight for those who have the opportunity to watch, and sound alone for those who prefer it for background entertainment.

In cities with more than one station, the audience will be divided if the several programs are telecast at the same time, but there is a possibility that agreements may be made whereby stations will complement each other's schedules to supply a large variety of entertainment to set owners. Local economics alone will work out details such as one station in each community providing all the television programs, or two or more outlets co-operating to the extent of scheduling their shows at nonconflicting hours.

While sponsors of television programs may disagree or be doubtful about various phases of present activity, most of them appear to be in accord on one vital principle that insures a healthy financial future for television. They agree on the greater power and appeal of sound-and-sight advertising over sound or sight alone. Instead of the advertiser's message being received by only one sense, that message will make a "double-barreled" impression on both ear and eye. "Advertisers are ready for television, but television isn't ready for advertisers" is a statement that possesses some truth, although it is gradually finding its way to the television museum.

Radio listeners would probably resent programs which

contained more than 10 to 15 per cent advertising. Yet they do not object to reading publications consisting of from 15 to 60 per cent advertising. At present it may be difficult to imagine a television program of which half would be direct advertising, but much indirect advertising may be used without annoying an audience. Many successful commercial motion pictures have used this indirect technique to good advantage, and once television develops a smooth technique for handling the "plugs," there is every reason to believe that it can do likewise. A sponsor's product, such as a car, can be brought into a script very naturally and yet be highly effective advertising.

There have been many speculations about the cost of television programs, some estimates scratching along rock bottom, others dealing in astronomical figures. A more realistic study of actual American and English programs indicates that while the cost of television entertainment may exceed radio, it need not be so expensive as motion pictures. The average class B film costs from $200,000 to $300,000—let us say $200,000. Radio programs vary widely in costs, reaching as much as $30,000 an hour for talent alone, but many fine programs have been built at a cost of from $5,000 to $10,000 an hour for talent. Television may find its proper niche somewhere above radio costs, but considerably under film costs.

Stations may share the cost of initial sustaining programs. Costs may be shared directly or indirectly through contracts with networks providing for programs. Present costs for a television hour range from about $1,000 to $2,500. Shortly after NBC started its regular telecasts in 1939, it was estimated that about $15,000 was spent weekly for a twelve-hour service, or $1,250 an hour. The various hours differed widely in cost, however, as commercial films were sometimes used without charge, while live talent shows with larger budgets were comparatively expensive. The lengthy rehearsals required for stage-type productions are expected to keep the costs of such shows at a high level.

While these figures are approximate, they present a general idea of television program costs. They are based on shows lacking the "name" value of stars. Better shows will, naturally, be more expensive. Show business, even with comparatively unknown players, is expensive. It is a completely unionized, closed-shop industry. Theatrical union officials, calling attention to the insecurity inherent in show business, insist upon high salaries with extra pay for rehearsals. Leaders of the various actor and musician unions have paid increasing attention to television during the last two years and are expected to set up special scales covering telecasts. Fair to very good wages are the rule in radio and film industries, and television will probably offer comparable wages.

Even doubling the maximum figure of $2,500 an hour for talent to provide for increasing costs as television develops, and allotting $5,000 for the television budget, the cost per station or sponsor if a limited network of only five stations were organized would be $1,000. Added to this would be time charges when this type of sponsorship is permitted. Complete figures are not yet available, but if this cost should average $1,000 a station, the total figure would be $10,000 an hour for time and talent, or $2,000 per station.

Radio advertisers figure from about one-third to one-half a cent as the cost of reaching each radio set. If there should be 200,000 homes—a reasonable figure—in each city looking at the telecast, it would cost advertisers one cent to reach each home. It is estimated that television will be at least three times more effective as an advertising medium than radio alone. It appears reasonable then, that a cost of two or three times more than present radio is practical for a medium that is at least three times as effective.

Another television service, one for theater use, is also a definite possibility. The way of theater television is not clear as yet, but it seems logical to expect a different type of service—otherwise it would offer little to attract patrons. Possibly certain sports events will be purchased exclusively by television-picture companies for showings at theaters only. When English sporting events were televised to theaters, seat sales boomed, and when final details are worked out the same interest may be expected here. Several private showings of theater-size television have attracted wide attention and approval, and special television programs for theater use appear to be a definite possibility.

Pioneer television advertisers will receive much valuable publicity. This will probably be of the word-of-mouth variety, as the amount of publicity newspaper publishers will grant to television is still doubtful. Naturally, they will carry news of genuine interest, but they will probably be hesitant about mentioning the names of television sponsors, feeling that by doing so they are building up a competitor. In some cases newspapers will be affiliated with television stations, offering greater opportunity for publicity. The publicity received by early radio advertisers was a valuable "bonus" for their pioneer and sometimes crude programs received on earphones or other primitive receivers.

Residents of the larger cities will undoubtedly be the first to whom television shows will be made available. The development of the comparatively inexpensive radio

The Finance Problem

relay system whereby television signals from one station may be picked up and "boosted" to the next station, thus dispensing with the more expensive coaxial cable method, will soon make network television programs a reality, although it may be some time before this service will be available for rural residents. It will be more practical to service first the concentrated markets, probably about 100 cities with populations of 100,000 or more. As about half the country's population resides in these communities, they are logical fans to receive television programs.

A nation-wide poll (American Institute of Public Opinion) conducted in the spring of 1939 revealed that one in every eight families considers itself a potential buyer of a television receiver. This would constitute a large initial audience, and it is probable that more families will join the parade once they see that their neighbors have television.

Television is developing its own particular technique slowly and painfully, just as after almost twenty years of vaudeville, stage, picture, and minstrel show technique, radio gradually developed its own entertainment form.

As television progresses it will probably be found that the best show is not necessarily the most expensive. On more than one occasion motion picture producers have been pleasantly surprised when what was regarded as a routine, class B picture did a surprisingly good business—possibly better than some class A pictures when the investments were taken into consideration. On the air, many popular and effective shows are not expensive. Low-cost quiz shows have been popular for several years and have done effective sales work for sponsors. They are neither expensive nor difficult to produce and often possess greater human interest than the more formal programs using professional casts. While the Major Bowes show is now in the upper brackets, it differs little from its early appearances on a local station at a very modest budget. In this case, its popularity earned larger fees, but its structure does not require an expensive production.

With the development of more expensive shows, new types of sponsorship may be discovered which will reduce costs. There are many "participation shows" on the radio in the daytime which have large followings. These programs are generally intended for the housewife and are often conducted by a popular woman commentator. Makers of noncompeting products purchase time or "announcements" on such shows. Their success indicates that such productions have definite entertainment value.

It is possible that a variation of this type of sponsorship might be followed on television programs. Makers of two noncompeting products, such as automobiles and watches, might jointly sponsor a program. The script could provide for bringing in "natural" mention or views of the products, such as close-ups of a family getting into a car, or Uncle Joe looking at his watch. The greater power of visual appeal should make for even greater effectiveness than that achieved by sound alone. Shows will probably be slanted toward the entire family as evening broadcasts will undoubtedly increase in attempting to reach the family group.

Television will be an important economic stimulus. Factories will require more skilled workers; trained sales staffs will be needed; entertainers, business executives, and employees will find new opportunities. Television will probably require considerable support for some time to come. Experiments, and expensive experiments they are, continue in the laboratories. Profits on sets will be small for the present time, and additional program services will increase costs. But because of the size of the investment already made in television, and because of its great potentialities for both service and profit, television must develop as a major force in this country.

CHAPTER ELEVEN

The Challenge of Television

BY ROBERT EDMOND JONES

WE are living in a period of extraordinary activity, of intense curiosity, of eager restless experimentation. Our time is a time of wonder. New forces are germinating. New forms are in the making. On every hand we are reaching out for new ideas, new ideals, new concepts. And—just as eagerly—we are reaching out for new ways in which to express them.

Perhaps the most striking example of this spirit of change is the astonishing development of all methods of communication in modern times. Many people now living once traveled from place to place in ox carts. Now these same people are shot through the air at three hundred miles an hour in the great new stratosphere liners. Not so long ago letters were delivered by post chaise. (Carefully penned epistles they were, filled with Spencerian flourishes and elegant turns of phrase.) But then came a strange new invention called the telegraph, the machine that could write at a distance. Then came the telephone and the radio, the machines that could talk at a distance. And now comes television, the invention that all but abolishes distance itself.

A television studio is one of the most exciting places to be found in the world today. Here is real creation. Something has been brought into the world that was not there before—a substance of things hoped for, an evidence of things unseen. As we sit in the tiny cramped control room, crowded with its elaborate mechanism, its complex switches and dials and telephones, watching the triplicate oblongs of quivering light out of which materialize the moving and speaking likenesses of the very actors we see on the stage before us beyond the soundproof glass panel, it is impossible not to feel that we are not only on the threshold of a great new medium of communication, but that we are about to set out on a great new world adventure. We simply do not know what the limits are to this voyage of discovery. Soon—very soon—we shall be able by pressing a button to enjoy the brilliant acting of Alfred Lunt and Lynn Fontanne in their current Broadway successes. Before long Bernard Shaw, sitting in his country home in England, will be able to attend the New York première of his latest play. Eugene O'Neill may witness the Theatre Guild's production of his new nine-drama cycle from his home in California. And this is only the beginning.

We Present Television

Behind the development of television there is a dynamic energy that is irresistible. It simply cannot be stopped. If every last fragment of research connected with this invention were destroyed tomorrow television would be invented all over again. What is the reason for all this expenditure of effort? Nobody asked for telephones or motion pictures. Even the radio was not a necessity. We got along well enough without these things. It is easy to account for the tremendous advances in medical research—treatments for cancer, for tuberculosis, for arthritis. There is a life-and-death necessity for the human race to attain this knowledge. But why the intensity of interest, the burning concentration on these new mechanisms that bring far things near?

Deep in us all is a longing to know one another and to be known. In the last analysis we are all lonely and alone, and in our loneliness we reach out to one another. The human urge to "get together" is ancient and strong. It is my belief that we are moving very rapidly toward a society in which every human being can be immediately present to every other human being in the world. Our lives are becoming more and more public. Privacy is being abolished. An unseen force is drawing us all together moment by moment. That force now manifests itself in television.

We cannot escape thoughts like these as we reflect

upon the thrilling potentialities of this new invention. We feel that we are living through one of the most dramatic moments in history. It is possible that at this moment we are witnessing the last step toward the creation of a new world unity, a world body in which we are destined to live and move and have our being. Could such a dream come true? We wonder. . . . But here is the instrument. Here is the hope. Here is the promise.

The essence of television is immediacy. The image we see on the television screen is not a painting or a photograph or even a motion picture. It is the representation of an event at the very instant of its happening. Motion pictures, of course, can be—and often are—broadcast by means of television. *But the quality peculiar to television is that of immediate presence.* An actor steps before the receiving machines in the studio and then and there presents himself to us in our homes—almost in the flesh, all but living. How different this is from a motion picture! When we see Garbo's current film, *Ninotchka*, we deceive ourselves into thinking that we are really seeing Garbo herself. After we leave the theater we will say, I saw Garbo in *Ninotchka*. But we didn't see Garbo at all. She made that picture months ago. We do not even know where she is at this moment. What we saw was a kind of magical memory of Garbo, as haunting and compelling as the memory of some long-lost loved one. Per-

haps that is why we call her performance *memorable*.
Perhaps the secret of motion pictures is that they haunt
us as memories haunt us. A well-known writer of motion
picture scenarios is quoted as saying, "You have to remember in writing film stories that the film audience is
not an audience that is awake, it is an audience that is
dreaming— It is not asleep but it is always dreaming."
The television audience, however, is wide awake, and television is neither a dream nor a memory. It is an immediate presence. It is an actor making a personal appearance all over the world at the same instant of time—a human being, just like you, just like me.

A dream is not reality. Yet all Americans worship the
movies. The great question as to whether television will
supplant the motion picture as a medium of entertainment resolves itself essentially into the question as to
whether we prefer realities to dreams or vice versa. Time
will answer this question for us. But we may note that
even Hollywood of late seems to be turning away from
its typical "escapist" films and giving its attention to such
stern presentations of fact as *The Grapes of Wrath*. This
motion picture seems real enough, agonizingly real at
times. Television, however, can give us a still deeper reality. In the not too distant future we shall be able to
watch the Joad family themselves as they make their ter-
rible trek from Oklahoma to California and they will live

out their bitter lives in our presence. Their very living will be transmitted to us and shared by us. We shall discover in them the qualities that make them like ourselves, the qualities of universal humanity that transcend all differences of appearance or custom or race or religion. Perhaps when we have become accustomed to seeing our fellow beings presented to us with such immediacy we may begin to understand them a little, and perhaps when we understand them we may even begin to like them.

At any rate, the public is interested not in televised movies but in people. It is true that the most popular form of television entertainment at the moment is the presentation of drama based on the conventions of motion picture technique. In fact, a televised drama today looks rather remarkably like a talking picture seen in little. The tiny screen seems to have a curious added vitality, the actors seem more real. That is all. Otherwise we might be looking at an unusually exciting "talkie" through reversed opera glasses. This may lead, as some think, to a new synthesis of theater and cinema. We may look forward to a single central repertory company whose performances are broadcast simultaneously all over the country after the manner of present-day radio concerts. The establishment of such a company is only a matter of time. We shall presently see extraordinary changes in the show business and in the motion picture industry. A new medium of dramatic expression is about to challenge them.

As I have said, the essence of this new medium is immediacy, a kind of supercandor. The images fairly leap from the screen, vivid and vital. Only color is needed to make them all but living. Presently color will come to television as it has come to motion pictures. And not long afterward, by some magical arrangement of curved mirrors and lenses and transparencies, the images will detach themselves from the screen and appear in space before us, with the stature and the colors and the voice of life itself.

These are the imaginings that fill our minds as we watch Jimmy Walker make his moving plea for the sufferers from infantile paralysis. His performance is uncannily effective. He comes near to us. He speaks to each of us personally. We know him more intimately than we have ever known him before. At the close we are in tears. Or Harry Hirshfeld draws a cartoon and signs it right before us. We see him create it. This is what is exciting, to watch the creative faculty at work. We seem somehow to get below the surface. We say, if we could only see Picasso do a painting!

The startling new intimacy between performer and public—more startling, more immediate almost than personal contact—doesn't make for art as we know it. We may be witnessing the beginning of a new form of art, but this art has yet to appear. The arts of literature, music, painting, are already established. They have been with us for a long, long time. But we have not yet been able to make up our minds as to whether photography is an art or whether motion pictures are an art, let alone whether television is an art. Certainly television has no more art in it at present than is involved in the presentation of a newsreel. And yet it presents an opportunity for the development of an art peculiar to itself, greater perhaps than any art we have ever known—an art so powerful that it may not even be governed by the laws of art as we have known them in the past.

I will give you an illustration. When we go to Mount Vernon, the home of George Washington, we see a carpet which Louis XVI of France had woven especially for Washington with the coat of arms of Washington in the center, and on the chimney piece a steel engraving of Louis XVI himself—his own likeness, a gift to Washington. That engraving was the best that King Louis could do at the time in the way of presenting himself to Washington. It was seen through the eyes of an artist, the best artist obtainable in France, and thereby it automatically became a work of art. It was idealized and ennobled—immortalized, as we say. Its resemblance to its subject was incidental, but it carried the idea of kingship, and in itself, as an engraving, it was beautiful. Now if Louis had lived in the Third Empire he would have sent Washington, not a painting or an engraving, but a daguerreotype. If he had lived in 1938 he might have sent a set of candid-camera "shots" of himself, or perhaps even several rolls of sixteen-millimeter film. And if by some miracle he were to return to earth in 1960 he would cross the ocean by television and present himself in Washington's drawing room.

You will observe that we get successively farther away from art—at least from visual art, as we have known it hitherto—as we get nearer to life. The old engraving was a product of careful reflection and painstaking execution. The new medium of direct instantaneous presentation gives little opportunity for selecting. The material is merely observed and projected.

The growing realism of photography, the candid camera, the motion picture, seems to have had the result of exposing people rather than revealing them. We have come of late to regard our fellow men as pretty poor creatures. The aspects of humanity that are shown to us do not credit humanity with its own inherent dignity. Our amazing new methods of communication have not succeeded in awakening in us a new sense of human truth, of the immensity and splendor of life. They seem instead to have given us brain fag, weariness, and an irritation with our fellows. Have we missed the point and the purpose of our great new inventions? We have the linotype, the electric organ, the talking picture, the radio, and now television. But we have no new Bach, no new Titian, no new Duse, no new Eugene O'Neill. We seem to have become conscious of a disconcerting and disillusioning averageness in life.

But the peculiar and special art of television will consist more and more in the appraising of truer and more fundamental values. We shall be given a selection of life, a proud display of exceptional people in exceptional situations. The artists of the future, working directly with life itself, will discriminate as sharply and as shrewdly as did the great artists of the past, and by their ruthless rejection of everything that is inferior will hold up to the world a new ideal of living. Here is the challenge of this new medium. Will some undiscovered genius lift our common experience into a higher region until we suddenly feel ourselves at one with the whole world? Or will television go the way of motion pictures and the radio? The producers seem to be far more interested today in promoting a television instrument in every home than in employing great dramatists, great actors and great painters to create in the new medium with insight and audacity. When will they realize that they must expend the same amount of energy and money in creating programs for the television apparatus as they expend on the development of the apparatus itself? They promise us a television instrument in every home. But what are we to see on that instrument?

Television so far has given us little beyond the sense of an extraordinary and very winning kind of novelty. We should not forget, however, that the industry is new. The pressure is great. Time is short. Money is limited. It is no wonder that most of the basic principles of the art of television remain undiscovered. But this is simply a fact of the moment. Presently television will mature into its own form, free and significant. Someone has said that true drama does not deal with the working of our minds but with the beating of our hearts. True drama does not concern itself with politics or religion or ethics or even logic. It does not preach or teach or illustrate. It only shows life. It says to us: Here is life. Look at it! Love it! If this is true—and it is true—then television is inherently dramatic.

Television is only waiting for the ones who will take it—now, in its beginnings—as a sculptor takes his block of marble, and will shape it into a new beauty and a new reality.

TELEVISION BROADCAST STATIONS

AS OF MAY 1, 1940

LICENSE AND LOCATION	CALL LETTERS
Columbia Broadcasting System, Inc. NEW YORK, NEW YORK	W2XAB
Allen B. Du Mont Laboratories, Inc. (AREA OF NEW YORK, NEW YORK)	W1OXKT
Allen B. Du Mont Laboratories, Inc. PASSAIC, NEW JERSEY CONSTRUCTION PERMIT, NEW YORK, NEW YORK	W2XVT
First National Television, Inc. KANSAS CITY, MISSOURI	W9XAL
General Electric Company BRIDGEPORT, CONNECTICUT	W1XA
General Electric Company NEW SCOTLAND, NEW YORK	W2XB
General Electric Company SCHENECTADY, NEW YORK	W2XD
General Electric Company SCHENECTADY, NEW YORK	W2XH
General Electric Company NEW SCOTLAND, NEW YORK	W2XI
General Television Corporation BOSTON, MASSACHUSETTS	W1XG

We Present Television

LICENSE AND LOCATION	CALL LETTERS
Don Lee Broadcasting System LOS ANGELES, CALIFORNIA CONSTRUCTION PERMIT, HOLLY- WOOD, CALIFORNIA	W6XAO
Don Lee Broadcasting System (AREA OF LOS ANGELES, CALIF.)	W6XDU
National Broadcasting Co., Inc. NEW YORK, NEW YORK	W2XBS
National Broadcasting Co., Inc. PORTABLE (CAMDEN, NEW JER- SEY, AND NEW YORK, NEW YORK)	W2XBT
Philco Radio and Television Corporation PHILADELPHIA, PENNSYLVANIA	W3XE
Philco Radio and Television Corporation PHILADELPHIA, PENNSYLVANIA	W3XP
Purdue University WEST LAFAYETTE, INDIANA	W9XG
Radio Pictures, Inc. LONG ISLAND CITY, NEW YORK	W2XDR
RCA Manufacturing Company, Inc. PORTABLE (CAMDEN, NEW JER- SEY)	W3XAD
RCA Manufacturing Company, Inc. CAMDEN, NEW JERSEY	W3XEP
State University of Iowa IOWA CITY, IOWA	W9XK

Television Broadcast Stations

LICENSE AND LOCATION	CALL LETTERS
State University of Iowa IOWA CITY, IOWA	W9XUI
Zenith Radio Corporation CHICAGO, ILLINOIS	W9XZV

Biographical Notes

CHARLES E. BUTTERFIELD dates his interest and experience in radio back to his boyhood hobby days. Formerly City Editor of the Champaign, Illinois, *News Gazette*, he has had all-round newspaper experience and bears the distinction of being the only newspaperman to interview Guglielmo Marconi via a two-way talking circuit between New York and Rome. Since 1927 he has been writing a column of news about radio for a nation-wide audience.

DONALD GLEN FINK is a graduate of the Massachusetts Institute of Technology, Class of 1933, with a Bachelor of Science degree in Electrical Communication. From 1933 to 1934 he was a Research Assistant in the Departments of Geology and Electrical Engineering of the Massachusetts Institute of Technology. He has been a lecturer on Electronics, Atomic Physics, and Radiant Energy for the Engineering Department of the Westinghouse Lamp Company. Since 1934 he has been on the editorial staff of *Electronics*, published by the McGraw-Hill Publishing Company, Inc., with the present position of Managing Editor.

He has contributed the Television section of the Third Edition of the *Radio Engineering Handbook*, and the section on Electronics and Radio Communication of the Seventh Edition of the *Standard Handbook for Electrical Engineers*, and is the author of the books: *Neon Signs* (with S. C. Miller), *Engineering Electronics*, and *Principles of Television Engineering*.

BENN HALL is on the editorial staff of *Radio Daily*, where he covers television, facsimile, and the commercial developments of the broadcasting industry. He is probably the country's first nontechnical television columnist, having contributed a weekly column of television business and entertainment news for *The Billboard*, 1932-1934. Mr. Hall later issued a confidential news-letter, *Television Times*, and has contributed articles on television to many publications, including the League of Nations' *Intercine* (published in five languages), the old *Review of Reviews*, *Literary Digest*, *Scholastic*, and others.

Mr. Hall attended New York University and the New School for Social Research. He served on the New York *Times* Sunday feature and radio departments, 1929-1936, when he resigned to become Associate Editor of *The Billboard*. He joined *Radio Daily* in 1938. Mr. Hall also contributes to various business publications and is the author of a book on business publications to be published this fall by Duell, Sloan, and Pearce, Inc.

O. B. HANSON, Vice-president and Chief Engineer of the National Broadcasting Company, has an international reputation as a radio engineer. Since the inception of radio broadcasting, he has contributed largely to its technical development. During a radio career spanning nearly three decades, Hanson was successively a radio operator at sea, chief testing engineer of the American Marconi Company, a pioneer broadcaster at Station WAAM, in Newark, New Jersey, and assistant to the plant engineer at Station WEAF, in New York City, before becoming associated with NBC.

When NBC was formed in 1926, Hanson went with the new company and since that time has directed technical operations and engineering activities for it. His association with television, of course, dates back to the days of mechanical scanning systems, through the period of experiments with ultra-short waves and the final installation of the all-electronic system at Radio City and the Empire State Building transmitter of NBC. At present he directs the activities of the entire technical staff of NBC television.

THOMAS H. HUTCHINSON graduated to television from the theater and radio. He began his theatrical career in 1913.

Biographical Notes

As actor and director, Mr. Hutchinson engaged in the production of plays in the larger cities of the United States. His writing experience includes numerous radio scripts and a comedy which achieved a Broadway run.

Hutchinson joined the National Broadcasting Company in 1928. After serving as program manager of NBC's Pacific Division, he left that position to represent a commercial account on the Pacific Coast. Hutchinson returned to NBC in 1934 as a program director. In 1937 he was made manager of the Television Program Division. As such, Hutchinson conducted many of the pioneer experiments in programming at Radio City during the years of experimental transmissions. When NBC began making arrangements for the inauguration of a public service in television, Hutchinson was charged with the responsibility of augmenting his staff of program directors, scenic designers, stage managers, and others to meet the demands of regular program transmissions. At present he supervises the production of every television program transmitted either from the Radio City studios or from field points by NBC's mobile television units.

ROBERT EDMOND JONES, one of America's outstanding stage designers, began his career in 1911. He designed the productions of *The Man Who Married a Dumb Wife, The Jest, Richard III, Macbeth, Redemption, Hamlet, Desire Under the Elms, Mourning Becomes Electra, The Lady With a Lamp, Green Pastures, The Passing Present, Night Over Taos, The Philadelphia Story,* and many others. He was associated with Kenneth MacGowan and Eugene O'Neill in the production of several plays at the Greenwich Village Playhouse from 1925, and has staged many of Eugene O'Neill's plays, as well as *Holiday* and *Mister Moneypenny* in 1928, *Serena Blandish,* 1929, and *Camille* in 1932. He began designing for color films in 1933, his work including *La Cucaracha* and *Becky Sharp.* He is the author (with Kenneth MacGowan) of *Continental Stagecraft.*

Biographical Notes

WALDEMAR KAEMPFFERT has devoted the last thirty years to the popularization of science and engineering. He was Editor of the *Scientific American* for eighteen years, and Editor of *Popular Science Monthly* for five years. He was the first director of the Museum of Science and Industry in Chicago, founded by Julius Rosenwald, and he laid out the plan which is now being followed in creating that institution. At present he is Science and Engineering Editor of the New York *Times*.

Mr. Kaempffert is the author of *The New Art of Flying*, which was one of the earliest books on aviation. His work naturally makes it necessary for him to keep abreast of the development of every branch of science. He is the author of *A Popular History of American Invention, Science Today and Tomorrow* (recently published), and of numerous popular articles on science, engineering, and industry which have appeared in the leading periodicals of the United States, England, France, and Germany. He writes all the editorials on science and engineering which appear in the New York *Times* and conducts for that paper, each Sunday, a weekly Science Department, which is used by many schools throughout the United States.

EARLE LARIMORE is particularly well qualified to represent the actor's attitude toward television, having had extensive experience on the stage, screen, radio, and now television. For eight years he was associated with the Theatre Guild, playing in *The Silver Cord, Strange Interlude, Marco's Millions, Volpone, Mourning Becomes Electra, Days Without End*, and many others. During the season of 1939-1940 he made a transcontinental tour in Ibsen repertory with Eva Le Gallienne. In radio, he has appeared in the Columbia Workshop and the Pulitzer Prize Plays series, *Alias Jimmy Valentine*, the serial *Life Can Be Beautiful*, and many other programs. In silent motion pictures he appeared in *Inspiration* and *The Kick-Off*. He appeared in NBC's first dramatic television production in the regular television program schedule, *The Unexpected*, and in the experimental television play, *May Eve*, and later *Confessional*.

HARRY R. LUBCKE received the degree of Bachelor of Science from the University of California at Berkeley in Electrical Engineering in 1929 and took graduate work in 1930. In November, 1930, the late Don Lee employed him to inaugurate the Don Lee television activities. As director of television he has been thus engaged for the past ten years in both program and technical operation and development. He is known by a number of patents held in the United States and foreign countries, as well as by contributions to the technical, trade, and artistic press.

ALFRED H. MORTON, vice-president in charge of television, National Broadcasting Company, is a radio engineer and executive of more than twenty years' experience. An engineering graduate of the University of Illinois, Morton began his career in the then infant industry of radio after resigning a captaincy in the United States Army in 1919. After two years in the service of the General Electric Company, he joined the Radio Corporation of America in 1921. First as manager of RCA's Washington office and subsequently as the corporation's European manager, Morton supervised the construction of pioneer radio stations in the United States, and at Rome and Milan in Italy. In the interim he was commercial manager of RCA Communications, Inc., for a period of six years.

Morton's introduction to the business and programming of broadcasting came in 1934, with his transfer to the National Broadcasting Company. He managed NBC's program department for three years and then became the chief of NBC's Managed and Operated Stations Division. When television was launched by NBC on May 1, 1939, Morton was named to head the country's first formal television broadcasting organization.

Biographical Notes

J. R. POPPELE, Chief Engineer and Secretary of Radio Station WOR, had his first encounter with radio at the age of fourteen, when he constructed and operated an amateur station with an early spark coil transmitter. He started his career during the first World War as a radio operator at sea and became WOR's engineer when that station first began broadcasting in 1922. He has since figured in the growth of WOR from its beginnings to the present 50,000-watt status. Once its only engineer, he now heads a staff of sixty technical experts. At the present time, Mr. Poppele is in charge of the station's engineering activities as well as its services in facsimile and frequency modulation.

JOHN PORTERFIELD, with a background of radio announcing and production, began his television experience in July, 1938, as director of one of this country's first experimental television program schedules for the National Television and Manufacturing Corporation. He organized and directed more than two hundred programs presented during a test period of a year to invited audiences who were individually queried on the content of each program. This is believed to have been the first series of television audience reaction tests given in this country and was reported in a series of articles published in *Cinema Progress*. He has appeared in many television programs for the National Broadcasting Company, in such varied assignments as an actor in the experimental play *May Eve*, program and mobile unit announcer, and master of ceremonies for variety programs. He has also directed commercial department store programs for the American Television Corporation.

KAY REYNOLDS began her writing career with the publication of poetry. Her work has appeared in the anthology, *The American Album of Poetry*, and in *Kaleidograph*, *Horizon*, and other poetry magazines. Her interest turning to the field of practical psychology, she published articles on this subject in the current popular psychology magazines. Her work with

Biographical Notes

various motion picture appreciation groups led to an interest in television, and a series of her articles on this subject was published in *Cinema Progress*.

THOMAS LYNE RILEY is a staff television director of the National Broadcasting Company. He has been connected with almost every branch of the entertainment business, with the addition of newspaper and other writing experience. He has been a magician, actor, and musician. In radio he was a writer, announcer, and director. He directed the first dramatic production in the regular NBC television program schedule, *The Unexpected*, on May 3, 1939. He has also directed the three comic operas, *Pirates of Penzance, Cox and Box*, and *H. M. S. Pinafore*; the plays, *May Eve, The Dover Road, The Farmer Takes a Wife, Post Road, The Perfect Alibi*; the first television program of one-act plays, including his own dramatization of Michael Arlen's short story, *When the Nightingale Sang in Berkeley Square*; the television performance of *When We Are Married* on March 3, 1940, the first current Broadway play to be televised in its entirety; a number of variety shows; the first Explorers' Club program, and the first television broadcast of grand opera, including a condensed version of the first act of *I Pagliacci*, on March 10, 1940.

Glossary of Terms

Amplitude: intensity of a signal.

Amplitude modulation: the changing in amplitude of a carrier wave correspondent to the variations in amplitude of the signal to be transmitted.

Audio (Latin: I hear): pertaining to the transmission of sound.

Bloom: glare caused by an object reflecting too much light into the lens of the camera.

Brightness control: the knob on the receiver which varies the average illumination of the reproduced image.

Camera tube: a device for converting light energy into corresponding electrical energy.

Carrier wave: the wave used for conveying the signals through the ether from transmitter to receiver.

Cathode: the electron source in a vacuum tube.

Cathode ray receiver tube: vacuum tube for converting electrical energy into corresponding light energy.

Center up: to center the composition of the picture at the television studio.

Close shot: a shot taken at close range, which includes a portion of the background.

Close-up: a shot taken at close range, which includes only the object or person televised.

Coaxial cable: a special type of cable suitable for conveying television signals.

Contrast control: the knob on the receiver for adjusting the range of brightness between high lights and shadows in a picture.

Control room: housing for the monitoring equipment from which the program is both directed and controlled.

Corn (slang): simple and obvious musical or dialogue arrangement in a program.

Glossary of Terms

Cycle: one complete alternation of an electric wave.

Dipole: a type of antenna for reception of high-frequency transmissions.

Dolly: a wheeled camera platform used to move the camera into different positions on the set.

Dolly shot: a shot taken when the camera is moving upon a dolly.

Electron gun: electron source for a strong and highly concentrated electron stream.

Facsimile transmission: the electrical transmission of a still picture, drawing, or document.

Fade-in: the gradual appearance of the screen image from total darkness to its full visibility.

Fade-out: the gradual disappearance of the screen image from its full brilliance to total darkness.

Focus: in the receiver, the adjustment of spot definition.

Frame: one complete picture. Thirty of these are shown in one second on a television screen.

Framing control: a knob, or knobs, on the receiver for centering and adjusting the height, width, and centering of the pictures.

Frequency: the number of cycles per second.

Frequency modulation: the changing in frequency of a carrier wave corresponding to the variations in amplitude of the signal to be transmitted.

Ghost: an additional and unwanted image appearing in a television picture as a result of signal reflection.

Gobo: a light-deflecting fin used to direct light in the studio and protect the camera lens from glare.

Hot light: a concentrated light used in the studio for emphasizing features and bringing out contours.

Iconoscope (slang: ike): a type of television camera tube developed by RCA.

Image dissector: a type of television camera tube developed by Farnsworth.

Interlacing: a technique of dividing each picture into two sets of lines, one set transmitted after the other, to eliminate flicker.

Glossary of Terms

Kinescope: a type of cathode ray receiver tube developed by RCA.

Line: a single line across a picture containing high lights, shadow, and half tones. 525 lines make a complete picture.

Long shot: an establishing shot taken from a distance sufficient to include a complete view of the scene.

Live talent: participants in a program picked up directly in the studio, as distinguished from film presentations.

Microphone boom: an adjustable crane which suspends the microphone.

Medium shot: a shot taken from middle distance, or from knee level to above the head of the subject.

Megacycle: when used as a unit of frequency, it is a million cycles per second.

Modulation: the process by which the amplitude, frequency, or phase of a wave is varied in accordance with a signal.

Mosaic: photosensitive plate mounted in the iconoscope and orthicon. The picture is imaged upon it and scanned by the electron gun.

Orthicon: a new, more sensitive television camera tube than the iconoscope and developed by RCA.

Panning: a horizontal sweep of the camera.

Polarization: the characteristic of the radiated carrier wave, e.g., vertical or horizontal, which indicates the direction of the electric field.

Process shot: a scene projected through a translucent screen which is televised as a background for action on the set.

Scanning: the action of the electron stream in exploring the mosaic in the camera tube or reproducing the elements of a picture on the fluorescent screen of the receiver tube.

Shading: reducing the undesired signals caused by the iconoscope in the process of scanning.

Signal: any form of intelligence transmitted by radio wave or wire communication.

Spot: the visible spot of light formed by the impact of the electron stream on the fluorescent screen of the receiver tube.

Glossary of Terms

Synchronization: the process of maintaining synchronism between the scanning motions of the electron streams in the camera tube and the cathode ray tube in the receiver.

Telecast: a television broadcast.

Telecine transmission: a program of motion pictures.

Television: the electrical transmission of a succession of images and their reception in such a way as to give a substantially continuous reproduction of the object or scene before the eye of a distant observer.

Tilting: a vertical sweep of the camera.

Trucking shot: a shot taken as the camera on a moving dolly transmits a scene.

Turkey (slang): a program that is a flop or failure.

Video (Latin: *I see*): pertaining to the transmission of transient visual images.

Womp: a sudden surge in signal strength resulting in a flare-up of light in the picture.

For Product Safety Concerns and Information please contact our EU
representative GPSR@taylorandfrancis.com Taylor & Francis Verlag GmbH,
Kaufingerstraße 24, 80331 München, Germany

Printed and bound by CPI Group (UK) Ltd, Croydon, CR0 4YY

01/05/2025

01858389-0009